EAT YOUR WAY
OUT OF CANCER

EAT YOUR WAY OUT OF CANCER

The Alternative Way to Healing
the Human Body Using Anti-Cancerous
Plant Foods and a Healthy Lifestyle

John Sammut

Copyright © John Sammut, 2015

All rights reserved. No part of this publication may be reproduced, stored in a retreival system, or transmitted, in any form or by any means, electronic, mechanical, photocopying, recording or otherwise, without the prior written permission of the author.

ISBN 978-1-56581-231-4

Cover photo credits:
Broccoli: Zsuzsanna
Tomatoes: Nota – freeimages.com

Medical Disclaimer

Before you read this book, I must give you the following warning disclaimer:

The medical information contained in this book is provided as an information resource only, and is not to be used or relied on for any diagnostic or treatment purposes. This information should not be used as a substitute for professional diagnosis and treatment.

Please consult your health care provider before making any healthcare decisions or for guidance about a specific medical condition. The author and publishers expressly disclaim responsibility, and shall have no liability, for any damages, loss, injury, or liability whatsoever suffered as a result of your reliance on the information contained in this book.

However, remember that if your country's health care system grants you no other option, but conventional health care treatment for cancer; you have the moral right, as a person suffering from any form of degenerative disease, to choose other forms of alternative therapies for the treatment of your disease. The final choice is always yours. I hope you seek and choose the right path of medical care.

Acknowledgements

Many people contributed to the making of this book. I would like to thank the following individuals who volunteered their time, talented efforts and by giving me their consent in making it possible.

Miss Elisabeth Patty – my girlfriend and future spouse

Dr. Michael Greger – written consent

Professor Jane A. Plant – written consent

Ty Bollinger – written consent

Not forgetting the many other clinical physicians, in-which the medical advice and information of some of this book was taken. I thank you all.

Contents

	Photo credits	v
	Introduction	1
1	What is the Key Diet for Homo Sapiens?	11
2	Cancer: Causes and Prevention	31
3	Breast Cancer	41
4	Making the Right Food Choices	53
5	When Good Men do Nothing!	59
6	What is a Plant-Based Diet?	65
7	The Disappointment of Conventional Treatments	75
8	The Importance of Choosing the Right Form of Vitamin B_{12} (Methylcobalamin)	85
9	The Health Benefits of Vitamin D_3	97
10	Strategies to Help Improve Local Healthy-Dietary Changes	107
11	The American Medical and Pharmaceutical Cartel	119
12	The War on Cancer	131
13	Medical Compliance with Lifestyle Medicine	139
14	Medical Malpractice	145
15	Guidelines on How to Deal with Cancer	153
16	The Health Benefits of Phytochemicals	163
17	The Cancer Environment	183
18	Supplementations, Juicing and Blending	203

19	The China Study with Reference to Breast Cancer	*213*
20	Mammography: Good or Bad?	*221*
21	Surgery, Chemotherapy and Radiation Treatments	*229*
22	Is there a Related-Link Between Wearing a Bra and Breast Cancer?	*239*
23	Genes do *not* Determine Disease on Their Own	*245*
24	Cow's Milk is the Perfect Food for Baby Calves. It is *not* a Healthy Food for Humans	*257*
25	Animal Products: The Slow-Growing, Fleshy-Road to Chronic Diseases	*291*
26	The Health Consequences of Pesticidal Residues	*341*
27	Carnism: Loving Some Animals and Eating Others	*349*
28	Rebounding: Cleansing and Boosting the Human Immune System Against Cancer	*361*
29	Best Anti-Cancer and Anti-Angiogenic Fruits and Vegetables	*369*
30	How Stress Influences Human Health	*393*
31	Transition to a Plant-Based Diet and the Significance of Detoxification for Improving Health	*401*
	Resources	*419*

Photo credits

page	credit
viii	Petr Kratochvil – all-free-download.com
10	Wikimages – all-free-download.com
18	Jason Gehrman – dreamstime stock photos
22	freefoodphotos.com
30	Inbarush68 – all-free-download.com
40	Andre Maritz – dreamstime stock photos
48	freefoodphotos.com
52	Vera Kratochvil – all-free-download.com
58	Scott Liddell – freeimages.com
64	Hans – all-free-download.com
74	Skeeze
79	Vera Kratochvil – all-free-download.com
84	Saulham
91	George Hodan – all-free-download.com
96	PixelAnarchy – all-free-download.com
100	Petr Kratochvil– all-free-download.com
106	Tpsdave – all-free-download.com
112	Petr Kratochvil – all-free-download.com
118	Geralt – all-free-download.com
130	Geralt – pixabay.com
138	PublicDomainPictures
144	Carlos Paes – freeimages.com
152	Bykst – ixabay.com
159	Niekverlaan – pixabay.com
156	Petr Kratochvil – all-free-download.com
162	Stux – all-free-download.com

182	PublicDomainPictures
192	freefoodphotos.com
202	theautomaticjuicer.com
207	HealthyFoodImages – pixabaycom
201	HealthyFoodImages – pixabaycom
212	Friendlymessage.deviantart.com
220	Magee-Womens Hospital of UPMC
228	Tpsdave – all-free-download.com
233	Tpsdave – all-free-download.com
234	Romanov – all-free-download.com
238	Mrharrison1 – all-free-download.com
241	Coloniera2 – freeimage.com
244	Epilepsyu.com
250	Totoley99 – all-free-download.com
256	Petr Kratochvil – all-free-download.com
262	Stux – all-free-download.com
267	Petr Kratochvil – all-free-download.com
272	Ikhangrafix – all-free-download.com
277	Petr Kratochvil – all-free-download.com
278	Peter Mazurk – pixabay.com
290	Hans – all-free-download.com
299	Hugh_Grant
317	freefoodphotos.com
322	Mike Johnson – freeimages.com
340	Tom Harpel – Wikicommons
348	Petr Kratochvil – all-free-download.com
354	Katzenspielzeug – all-free-download.com
360	lumc.org.au
364	bargainoutfitters.com
368	Jackie Egginton – dreamstime stock photos
376	Zsuzsanna

380	freefoodphotos.com
392	George Hodan – all-free-download.com
400	Chrizzel_lu – all-free-download.com
410	Byrev – all-free-download.com
418	Jarmoluk

Introduction

During the past fifty years scientists experimenting with thousands of animals have found 700 ways of causing cancer. But they have not discovered one way of curing the disease.

J. F. Brailsford, British M.D., M.R.C.S., Hon. Radiologist

As an experienced Food Bacteriologist, my specialization is food and water safety. However, my interest in health studies goes far beyond food and water safety. I am also a qualified Nutritionist and a Vegetarian and Vegan Consultant. I have always been fascinated by the study of human pathology, especially focusing on chronic-related diseases in our species. Why are diseases with us? After many long years of study and trying to figure out the correct human diet and its cancer connection, the answer has become apparent to me and to many other clinical medical scientists around the world.

This book is mainly focused and explains the correlation between diet and disease, especially cancer. Breast Cancer and all other types of cancers are all associated with an increased incidence risk from foods of animal origin. This book also explains the impact that the typical Rich-Westernized Diet has on the overall ecosystem, on animal welfare and towards the relationship and link between human health, and disease. Likewise, the opposite is always reported in clinical trials, whenever a plant-based diet and other healthy-lifestyle changes are examined – the ecosystem is adequately maintained, animal welfare is respected and human health is optimized.

This book also acknowledges and tries to unravel the unethical facts about the medical/drug cartel and the cancer industry political and financial burdens on human health. Strangely enough, whenever a real cure for cancer is found, large billion-dollar food and drug industries, including the cancer industry always attack and have negative views about a natural remedy or intervention plant food: that helps towards prevention and the reversal of cancer. The may or so-called cure is either dismissed or locked-up away from the general public's view, and may never be patented.

Importantly, I will also focus on the strategies of how you can reduce your own risk of developing breast cancer: the natural preventive treatments and risk factors involved (some of which have not been fully promoted, empowered and employed by governments and a large number of health-care systems). Finally, and more importantly I will explicate about how you can reduce your own risk of disease by following a Dietary "Cancer Management Plan" by using lifestyle medicine, which includes:

INTRODUCTION

1. Specific anti-cancerous plant foods, incorporated with,
2. Moderate amounts of daily exercise.
3. Stress management, and other varieties of non-drug modalities.

The conclusions and the statistical data contained in this book are based on, and refer to, studies carried out mainly in the United States, including many other countries. Nevertheless, the conclusions derived therefrom can also be applied to my own present country of residence, which is Malta, and to the whole Western world (namely European and all other countries), that have followed American dietary patterns and that, consequently, have developed chronic-degenerative diseases.

Studies have shown that non-communicable diseases (also known as chronic diseases), have indeed, dramatically increased: within populations that consume fewer amounts of plant-based foods and consume more junk foods: Animal-based products, processed and other unhealthy low-nutrient dense refined foods.

Does adherence to the *World Cancer Research Fund/American Institute for Cancer Research guidelines* actually reduce our risk of cancer? If, basing ourselves on the results of the largest "Prospective study on diet and cancer in history"[1] we succeed in managing our weight, eat more plant foods and less animal foods, consume less amounts of alcohol, and start life with mother's own breast-milk, we *can* significantly lower our risk of breast cancer and of all cancers combined.

Studies have shown that eating mostly foods of plant origin appears to be the most powerful and most protective tool for the prevention of chronic diseases. For example, a study carried out in the UK[2] found that in just one year there were 14,902 cancer cases caused by 'something' to which participants were exposed 10 years earlier. What was that 'something' that ended up causing thousands of cancers? That 'something' was an inadequate intake of fruit and vegetables.

If instead there *was* a toxic chemical spill causing almost 15,000 cancers, I am sure that people would be up in arms to ban it. But instead, when that primary killer factor is 'just' *not* eating their "fruits and vegetables" (as the Brits would say), it hardly gets anyone's attention.

Researchers have created a healthy lifestyle index, defined by four things:[3]

1. Exercise.
2. A dietary shift away from the standard American diet that is high in meat, dairy, fat and refined sugars, towards a more prudent dietary pattern, for instance, consuming more green and yellow vegetables, fruits and beans.
3. Avoidance of tobacco, and
4. Avoidance of alcohol.

Young women scoring higher on these four things cut their odds of getting breast cancer in half; older women cut their odds of breast cancer by 80%!

Clinical scientific evidence has clearly proved that lowering your risk of cancer and the prevention of all diseases can be retained simply through diet and lifestyle changes.

Most persons get half a dozen serious episodes of cancer during their lifetime without even knowing it! The human body can heal cancer on its own. This can be achieved by a well-functioning immune system. A good immune system keeps cancer at bay and reverses cancer episodes. On the other hand, in the long-term, an unhealthy immune system can never achieve this.[4]

Cancer is a long-term progressive disease which takes many years to develop, but once it does progress it becomes more difficult to control or reverse. Basically, the sooner you adopt a powerful anti-cancer diet, the better your chances will be to reverse cancer and avoid any relapses. Dietary reversals have occurred in all stages of cancer, even in cases of patients who were given only a few months to live. Indeed, one should never give up hope of curing oneself of this disease!

The plant-based diet as outlined in this book is strict (absolutely imperative), as it has to be, especially if you have already been clinically diagnosed with a serious illness – the more you understand it – the closer you adhere to it – the more effective a plant-based diet will be.

INTRODUCTION

> **FACT BOX**
>
> **SUGAR (contained in refined/processed foods and beverages) is FORBIDDEN as "Sugar Feeds Cancer."**

One can intelligently understand that the collective medical expertise and knowledge of many professional clinical scientists on disease prevention and disease reversal, from which this book derives, is far more successful and superior, than the medical expertise and knowledge of a single person.

Although this book's main focus is on combatting breast cancer, the same Dietary Cancer Management Plan, using lifestyle medicine, may also help in the reversal of many other forms of degenerative diseases.

Once you study this book you will comprehend that sugar is basically the main energy source for cancer: therefore food sources containing high-carbohydrate contents, including saturated fats and cholesterol, should be avoided!

Once cancer is under control one can return to a complete plant-based diet, containing wider food choices comprising adequate amounts of natural sugars: a variety of fruits containing healthy nutrients, antioxidants and more complex carbohydrates. This book offers you the vital information that is essential to help you achieve the above lifestyle changes in treating yourself from chronic disease. Remember that in the case of many people, it is not the disease that is killing them, but their bad diet and lifestyle.

There is only one option for cancer patients using conventional treatment: the three protocols of Slash, Poison and Burn: commonly known as surgery, chemotherapy and radiation.

Most certainly these medical procedures can extensively damage the body's immune defences and can lead to organ failure, secondary tumours and, sometimes, death. The usage of natural, non-harmful alternative treatments is the healthier, more advanced and more sensible option and which, ultimately, can counter-attack cancer. Natural treatments will boost your health and not bring forth harmful side-effects to the human body.

A "plant-based diet and a change in lifestyle" more effectively and most certainly has extremely powerful and synergetic anti-cancer effects, and which help combat the fight against this disease. Cancer has been acknowledged and recognized by many medical scientists and the general population to have increased in the past 100 years.

Almost every popular magazine features nutrition advice: Articles in newspapers, TV and radio programmes constantly discuss diet and health issues. Given the vast amount of information available, are you indeed confident that you know what you should be doing to improve your health?

If you are not completely sure of your answer, you may rest assured that you are not alone. In spite of the vast amount of information and opinions that are available, *very few people know exactly what types of foods and lifestyles factors they should be intervening to improve their health.*

We may not need to take drugs, or undergo unnecessary surgical procedures, or employ other toxic chemotherapeutic conventional treatments for all known common chronic diseases. Choosing alternative treatments is the best way forward. Lifestyle medicine will also greatly boost a person's health naturally, and will help render the body's immune system protective against any form of chronic disease. This is where we should be directly focusing upon, if we want to tackle these pandemic diseases in Westernized fast-food nations.

Here are some of the questions that we need to answer first:

- Are human's carnivores, omnivores or natural herbivores?
- Is diet the most essential and effective preventive strategy for disease prevention?
- Are industrial carcinogens the primary cause of cancer?

- Why is lifestyle medicine the key cancer management plan for disease prevention?
- Are conventional chemotherapy, radiation and surgical medical procedures beneficial and necessary for the treatment of cancer?
- Should you buy organic labelled food items to avoid pesticide exposure?
- Is your health "predetermined" by the genes that you inherited when you were born?
- What vitamins, if any, should you be taking for the prevention and treatment of cancer?
- Should you be eating animal-based products, and if so, how often?
- Is there anti-cancer foods (plants) that can help reduce cancer growth, which work just as effectively as, or better than conventional medicine, without producing any side-effects to the body?
- Why cancer?
- Is cancer an irreversible disease?
- What is the cancer environment?
- What essential nutrients are missing in a plant-based diet and why?
- Can humans get all the essential nutrients, including adequate amounts of calcium and protein on a vegetarian or vegan diet?
- What plant foods work best to help reduce cancer growth?
- How can you transition to a plant-based diet, and why should you?

By thoroughly reading and following the guidelines contained within the pages of this book, and evidently complying to these corrective measures you **will** find that your health will slowly improve. Scientific data has long proved that diet is the leading cause of chronic-related diseases in Westernized populations.[5]

We need to reboot our health-care system and start empowering out the right information and educating the general public, doctors, nurses, and physicians, etc., about the long-term health benefits of lifestyle medicine.

Medical tactics that do not use surgical procedures, harmful medicines and symptomatic medical approaches to diagnose, and treat only patients' symptoms: whilst leaving out the most important aspect of health – which is treating the underlying causes of chronic diseases. Today's conventional medical interventions will *never* be as valuable or effective as lifestyle medicine, for the treatment of chronic diseases.

A healthy preventative action plan is desperately needed.[6] However, many medical scientists are now realizing that lifestyle medicine is the primary imperative and the acceptable strategy needed to accomplish this task. Conventional medicine needs to adopt a new approach towards disease prevention and treatment. Again, as in most cases, the underlying causes of disease are left unresolved.

Conventional medicine contribution to "acute disease", has been fully acknowledged, mastered and accomplished spectacular health achievements in the 21st century: from treating bacteria, viral and parasitic manifestations, broken bones, cuts, burns and other trauma incidences, etc. However, "allopathic medicine" will never succeed in eradicating "Chronic-Related Diseases" from society: unless it emphasizes and focuses on lifestyle medicine. I truly am confident that readers of this book will realize that lifestyle medicine is the best way forward to reduce mankind's risk of non-communicable diseases,[7] including the major killers in Westernized populations: diabetes, obesity, autoimmune diseases, heart-related diseases, strokes and cancers.

Lifestyle medicine is the best medical treatment for the prevention, control and reversal of many forms of degenerative and progressive diseases: a plant-based diet in combination with other long-term effectiveness lifestyle interventions.

So, do try to look after your health, your family's health, animal welfare and the health of the overall ecosystem. All these important health aspects can be accomplished and resolved by just changing your daily diet and lifestyle – from an omnivorous diet to a complete herbivorous, plant-based diet, including other healthy-lifestyle interventions.

Notes for introduction

1. Romaguera, D *et al.* (July, 2012). Is concordance with World Cancer Research Fund/American Institute for Cancer Research guidelines for cancer prevention related to subsequent risk of cancer? Results from the EPIC study. *The American Journal of Clinical Nutrition*, **96**(1), pp.96:150-163. doi: 10.3945/ajcn.111.031674.

2. Parkin, D.M *et al.* (2011). Cancers attributable to Dietary factors in the UK in 2010. *British Journal of Cancer*, **105**, pp. S19-S23.

3. L.M. Sanchez-Zamorano *et al.* (May, 2011). *Healthy lifestyle on the risk of breast cancer. Cancer Epidemiol Biomarkers Preview*, **20**(5), pp.912-922. Greger, M. (2014). *Breast Cancer and Wine.* Retrieved from http://nutritionfacts.org/2014/03/20/breast-cancer-and-wine/

4. Quillin, P. (May 20th, 2005). Beating Cancer with Nutrition. *Nutrition Times Press* (4th Edition), pp.1-21.

5. National Cancer Institute (1985). Cancer Rates and Risks. Washington. R. Doll & R. Peto (1981). The Causes of Cancer: Quantitative Estimates of Avoidable Risks of cancer on the United States Today. *Journal of the National Cancer Institute*, **66**, pp.1191-1208.

6. WHO (2008-2013). Action Plan for the Global Strategy for the Prevention and Control of Non-Communicable Diseases, pp.1-42.

7. C. Roberts & R.J. Barnard. (2005). Effects of exercise and diet on chronic disease. *Journal of Applied Physiology*, **98**(1), pp.3-30.

1

What is the Key Diet for Homo Sapiens?

And God said, behold, I have given you every herb bearing seed, which is upon the face of the earth, and every tree, in the which is the fruit of a tree yielding seed; to you it shall be for meat.

Genesis 1:29
King James Bible "Authorized version", Cambridge Edition.

In the Beginning and the Formation of the Solar System

The "Big Bang Theory" is still the scientific theory of how the Universe started. A phemonoum, now accepted by many astronomers and cosomolgist's and billions of people around the world. A rapid inflation of an incredibly small pinpoint of pure energy. A "singularity" that expanded exponentially space-time, and then slowly cooled. As this happened, it changed energy into matter producing the atoms that went on to form the known Universe.

Over billions of years, heat and pressure within the Universe formed particles of matter and anti-matter. As this immense heat and pressure began cooling, matter and anti-matter violently collided and cancelled each other out – which only left a faction of matter to make up the entire universe. Any sort of explosion or expansion has nothing to do with the shape of bodies in space. This is controlled by gravity. As the mass of an object (the amount of stuff it contains) increases so does its gravity. When the object reaches a certain mass its gravity becomes so strong that it pulls all the matter in the most stable shape – clustered around the center point in a sphere, forming the constellations and planets.

As time slowly passed the Universe cooled and expanded, this created small amounts of matter to take the form of gases, called "nebulas". In these nebulas, so-called black holes, quasars, clusters of galaxies and other galaxies transpired and transformed: including the Stars, planets, moons and other planetesimal/celestial objects, that indeed, can be clearly seen and sometimes detected in the known universe. How the Solar System and organic life were formed and started on planet Earth is still a mystery, but many postulations have been theorized. The most accepted theory being:

A protostar, similar to the Sun exploded around 4.5 billion years ago. It was from its super-nova remnants that all known matter within the Solar System – asteroids, comets, meteoroids, planets, moons and the star we call the "Sun" were created.

The Earth's "Giant-Impact Hypothesis"

The early solar system was a violent place, and a number of bodies were created that never made it to full planetary status. According to the giant impact hypothesis, one of these celestial bodies collided with the Earth: not long after the young planet was created. A Mars-size body (Theia) collided and shattered the earth and shaped the earth and its satellite we know of today (the Moon). Maybe, in this collision life may have found the right conditions to help set off the biochemical-tree of life.

Abiogenesis:
The Natural Process of Life Arising from Non-Living Matter

Scientific hypotheses about the origins of life can be divided into three main stages: the geophysical, the chemical and the biological.[1] Many approaches explore how self-replicating molecules or their components came into being. On the hypothesis that life originated spontaneously on Earth, the Miller–Urey experiment and similar experiments demonstrated that mostamino acids, basic chemicals of life, can be racemically synthesized in conditions which were intended to be similar to those of the early Earth. Several mechanisms have been investigated, including lightning and radiation. Other approaches ("metabolism first" hypotheses) focus on understanding how catalysis in chemical systems in the early Earth might have provided the precursor molecules necessary for self-replication.[2]

Single cell organisms, like algae, fungi, bacteria and viruses may have been the first known microorganisms to evolve on planet Earth. Over billions of years other complex life-forms adapted to Earths environmental conditions and found a way to multiply and grow into more intelligent and advance species: bringing forth the millions of species we know exist today. These include all the animal and plant kingdom, insects and other life-forms, including primates and Homo sapiens (Modern Man).

We live on a so-called fragile planet (the Earth) and so we should admire and protect it, because it is the only planet we have at this present moment in time, which is known to be able to sustain organic life. Humans have been given a chance to explore, feel, create, evolve and learn the mysteries the universe has to offer us. We should unquestionably take great advantage of it.

It is a rare phenomenon in the Universe, where a planet (the Earth) is positioned at the right distance from its rotating star (the Sun). The Sun releases a constant amount of cosmic energy that helps start the fuel of biological metabolic reactions towards bacterial/plant/plankton photosynthesis, etc., that in a significant way starts the food-chain and cycle of organic life.

The moon (Earth's only satellite) which, even without us sometimes noticing, rotates around us, also provides and brings a balance to the Earth's weather and seasonal cycles of life. Let us all be grateful and be more respectful towards the Sun, the moon and the Earth for their contribution in achieving the correct balance and for the beneficial sustenance they provide to all species of life on Earth.

All this, which most humans take for granted, occurs every second, every minute, every hour, day in day out, year after year in Earth's history and in our own lives. We should look deeper into the cycle of life and towards our planet's future: because, alas, in a split second and in the blinking of an eye, all human life and all life's wonders could disappear faster than they have evolved.

Although actually a few million years old, the advent of human life on Earth is relatively recent. If we were to compare the Earth's evolution history to a 24-hour clock, the human species would be only a few minutes old!

We are just a diminutive, infinitesimal speck of dust or sand on a long stretch of beach. We are insignificant when compared to the complex reality and the gargantuan evolutionary tree of the universe. We are so-called "unique and special" because we think we are the most intelligent life-form in the Universe. This could be the truth, however, if this is not so, that is, if there are other and more advanced, intelligent extraterrestrial beings within the Milky Way or in other galaxies, then our way of thinking about life would definitely change. If this is the case, then our place and expectations in the cosmos would be dramatically influenced by these new postulations.

We are complex, intricate and convoluted beings with great minds and intelligence. Our minds wander towards reaching great heights with multiple expectations. There is so much we want to accomplish during our lifetime. Let us do the ultimate humane thing, that is, to respect all animals and other species on planet Earth and, maybe, by any means attempt to achieve and actually give something positive back to Earth, before we finally close our eyes and become once again bonded with this Universe.

In this macrocosm we were born, and from the Universe we will fade away and die. What I believe is this: what we leave behind us will be the most perceptible aspect of life. What we accomplish and give back to the universe will, maybe someday, abet others to better understand the connotation of the creation of this complex and mysterious universe.

Life after death is still a mystery to us. We do so much desire and hope that there is a God: a superior being who can hear our voice and read our thoughts. A God that can feel empathy and compassion towards all life forms in the universe. A God who understands human emotions and acknowledges our pains and sorrows, and, ultimately, recognizes our attempts to understand the complexity of life. A God who truly and fully understands that we desperately need to venture on after death. Unquestionably, we are special beings that are capable of showing love, expressing other strong emotions and achieving miraculous feats. Should that matter?

We live in a Universe, and so we have no choice but to contribute towards its destruction or amelioration. The Universe main focus is not on life: not the galaxies, planets, suns or moons or any intelligent life-form that may well cross its path. The universe is the master of all life, but in a certain sense, life is the master of the universe. The universe may need us, but we need the universe. The concept of life is life itself. Without life the universe may not have no significance. There should be no excuse. We should not rely on any God, but on ourselves. With or without a Creator, we should still cherish the precious time we have: in achieving our goals, respecting all other species that share this planet with us, genuflect before this amazing blue planet, and be united with each other by showing compassion, respect, love, devotion and accomplishing the best we possibly can, before we reach the end of our lives.

There are so many questions that need to be answered, but questions we should take seriously in order to understand where our place in life is. It is important to try to understand where we all might be heading in this spectacularly amazing, mysterious and wonderful, but undoubtedly, also chaotic, colossal, forceful, dangerous, violent, destructive and complex reality we have so called the Universe or Cosmos. Are human beings so special?

As the late astrophysicist Carl Sagan, whom I admire dearly, for his brilliance and knowledge about cosmology, rightly and intelligently quoted:

"We are made of star stuff."

"You have to know the past to understand the present."

"We are butterflies that flutter for a day and think it's forever."

"We are the product of 4.5 billion years of fortuitous, slow, biological evolution. There is no reason to think that the evolutionary process has stopped. Man is a transitional animal. He is not the climax of creation."[3]

People throughout the earth had always thought that they were at the hub of the Solar System until Copernicus found that after all it is the Sun that is the center, and that all other planets orbit around it. Columbus's failure to fall off the earth's edge changed the flat map into a globe, whilst Newtonian principles relating to velocity, acceleration, energy and mass forged the structure of physics. Einstein's theory of relativity shook the very foundations of physics, and all of mankind should not forget Stephen Hawking's, Michio Kaku's and the late Carl Sagan's, including many other astrophysicist's contributions: concerning the laws of physics and cosmology.

The World of Nutrition

Similar advances have been made in the world of nutrition. For centuries we thought of food choices as a modest force of medicine. But we were wrong. In nutrition, an optimistic and dynamic world has replaced the flat vista of the past.

We used to believe that heart disease can only be prevented and not reversed. But now medical scientists know that with the proper diet and lifestyle changes, heart disease can actually be reversed without drastic operation procedures and drugs.

We now have a new arsenal in the war against cancer. We can actually control our own fate and factors in controlling cancer. Dietary factors can alter hormonal levels, boost immunity, and improve the odds of survival of a cancer patient. Diet and lifestyle changes can even reverse, control and often cure many forms of diseases.

A change in eating habits has already begun. Millions of people worldwide who have become vegetarians or vegans have shown that we can live healthier lives without the consumption of animal produce. We now know that plant-based foods protect us from disease, while epidemiological studies have shown us that animal-based foods increase our risk of disease. This is, actually, old knowledge.

If we make the right food choices, we will enjoy longevity and self-control health. For many, it can mean a higher level of health we did not think truly possible. A plant-based diet and other lifestyle changes can help us accomplish this.

The Evolutionary Diet of Primates

Primates are believed to have evolved in tropical forests. Even today, tropical forests are where most primate species are found. Indeed,

the most recent paleontological evidence suggests that the earliest known hominid (*Ardipithecus ramidus*, a taxon estimated to be some 4.4 million years old) lived in a wooden rather than in a more open savanna environment.[4] According to Katherine Milton,[5] primates, being forest dwellers, have found the foods available in most of their evolutionary history that is, leaves, fruits, flowers of tropical forest trees and vines.

Almost without exception, extant apes and monkeys take the greatest proportion of their daily diet from plant foods – new leaves, ripe fruits, seeds, exudates, nectars, flowers, pith – eating only moderately to trace amounts of animal matter, generally invertebrates (Milton, K. 1999).

All great apes are remarkably herbivorous. Gorillas and orangutans are estimated to acquire most of their annual diet from plant foods.

Primates only eat minuscule amounts of animal matter, largely invertebrates. Although chimpanzees "fish" for termites and "dip" for ants as well as hunt and eat vertebrate prey (often small monkeys), such foods tend to make up only a minute percentage (4-6%) of their main diet: mainly composed of leaves and of large ripe fruits.[6]

Using data from various lines of evidence – anatomic, paleontological, and physiological – there seems to be general agreement that the ancestral line leading to apes and humans was remarkably herbivorous.[7]

If humans eat animal products, the gastrointestinal track (GIT) transit-time takes up to 2-5 days to fully digest and break down corporeal flesh, whilst plant foods that contain plant roughage (fibre) may take only 12-48hrs. A longer GIT food transit time is a health problem. The extensive time period needed for the human GIT to adequately break down animal flesh can lead to the accumulation of toxic build-up in the body. Without the essential micronutrients and fibre, which indeed, are found mostly in plant foods – toxins, carcinogens, excess cholesterol and growth hormones can cause a toxic-overload environment in the GIT. Long-term consumption of animal flesh may eventually lead to the development of chronic colon complications and other systematic health problems in humans, including the risk of developing hormone-dependent cancers.

Medical evidence overwhelmingly authenticates that the more we consume animal products, and the less amount of plant vegetation we eat, the more sick we will get: heart disease, cancer, diabetes, osteoporosis and many other chronic degenerative diseases. If eating meat were so natural, it would not destroy our health.[8]

Are Human Beings Herbivores, Carnivores or Omnivores?

Although most of us conduct our lives as omnivores – that is, we eat animal flesh as well as plant vegetation – human beings have characteristics (anatomical and physiological features) that are closer to herbivores than to omnivores, and especially not to carnivores. How do we know this? Let us have a look at some different features between carnivores and herbivores below, as prepared by Dr. William C. Roberts, editor of the American Journal of Cardiology:

1. The intestinal tract of carnivores is short (3 times body length); that of herbivores, long (12 times body length).
2. Body cooling of carnivores is done by panting; herbivores, by sweating.
3. Carnivores drink fluids by lapping; herbivores, by sipping.
4. Carnivores produce their own vitamin C; whereas herbivores can only obtain it from their diet.
5. The teeth of carnivores are sharp; those of herbivores are mainly flat (used for grinding or biting into edible fruits).
6. The appendages of carnivores are claws; those of herbivores are hands or hooves.
7. The tongue of carnivores is rough and thin; in herbivores it is smooth and thick to manipulate vegetable matter into the back molars for grinding.
8. In carnivores their blood and urine is acidic; the blood and urine of herbivores and humans is alkaline.
9. The Animal Liver: in relative terms, a carnivore's liver is a tool designed with the capacity to eliminate ten times as much uric acid as the liver of man or other plant eater.

In a 1991 editorial in the American Journal of Cardiology, Dr. William C. Roberts wrote:

> Although we think we are one and we act as if we are one, human beings are not natural carnivores. When we kill animals to eat them, they end up killing us, because their flesh, which contains cholesterol and saturated fat, was never intended for human beings who are natural herbivores.[9]

In the late 1940s, Western physicians went to Africa and made a very surprising discovery about the health of the population they studied. They noticed that many of the diseases seen in the Western world were not present; they were not seen in the African population that were living particularly in the rural areas. They didn't see, for example, atherosclerosis, coronary artery disease, blockages (and) hardening of the arteries. They did not see high blood pressure. They did not see appendicitis, cholecystitis, gallstones, kidney stones, (and) our three most common cancers: cancer of the breast, colon and cancer of the prostate gland. They did not see polyps in the colon or hemorrhoids. Even osteoarthritis and osteoporosis were very uncommon."[10]

In this context, it is worth noting Dr. W.C. Roberts' comment, namely:

> There were a number of articles in medical journals and medical books published in the late 1940s, 1950s, from Africa, talking about the 'Western disease', the 'diseases of progress', the diseases that were killing people in Europe, in the U.S.A, and in all the countries which were eating cows, pigs, chickens, sheep, and goats with every meal.

The Correct "Human Diet"

As intriguing these arguments may be, the idea that humans are natural vegetarians has "no scientific basis, in fact," argues anatomist and primatologist John McArdle. Alarmed by this growing belief, McArdle, a vegetarian, says that human anatomy proves that people are omnivores.[12]

John McArdle, executive director of the Alternative Research and Development Foundation in Eden Prairie, Minnesota, states: "We obviously are not carnivores, but we are equally obviously not strict vegetarians, if you carefully examine the anatomical, physiological and fossil evidence."

There have been many debates about this argument. According to a 1999 article in *The Ecologist* several Homo sapiens physiological features "clearly indicate a design" for eating meat, including our stomach's production of hydrochloric acid, sometimes not found in herbivores. Furthermore, the human pancreas does manufacture a full range of digestive enzymes to handle a wide variety of protein foods, both animal and vegetables. The same article in *The Ecologist* continues to say that: "While humans may have longer intestines than animal carnivores, they are not as long as herbivores; nor do we possess multiple stomachs like many herbivores; nor do we chew cud.

Our physiology definitely indicates a mixed feeder."[13]

Nevertheless, Dr. Neal Barnard, founder and president of the Physicians Committee for Responsible Medicine (PCRM), says that humans lack the raw abilities to be good hunters: "We are not quick, like cats, hawks, or other predators. It was not until the advent of arrowheads, hatchets and other implements that killing and capturing prey became truly possible."[14]

Even though there is a great deal of debate going on about dietary issues on whether or not we are omnivores or completely herbivores in nature, the scientific evidence regarding human health cannot be ignored.

Whenever we drink or eat any form of animal foods the more our risk of developing disease increases. This means that all humans should be primary focusing on eating mostly plant foods and very little amounts of animal flesh. Otherwise we will suffer the consequences of heart disease, cancer and other degenerative diseases.

William J. Mayo, founder of the famous Mayo Cancer Clinic in the United States, addressed the American College of Surgeons with these words:

> Meat-eating has increased 400% in the last 100 years. Cancer of the stomach forms nearly one-third of all cancers of the human body. If flesh foods are not fully broken up, decomposition results, and active poisons are thrown into an organ not intended for their reception.[15]

In his book, *Health Wars*, Phillip Day wrote as follows:

> We chomp meat because our society was brought into the fear of dying through lack of protein, most believing that unless we ingest the animal flesh, we are in serious danger of becoming protein-deficient (kwashiorkor or marasmus). This myth originated from

early trials conducted on rats. Later, it would transpire that rats require up to eleven times more protein than humans, as evidenced by the commensurate increase in rat mothers' proteins in milk as compared to the protein content of human milk. Today, it is recognized that protein requirements are not nearly as big as formerly assumed. Nevertheless, the protein-scoffing head has been hard to exorcise from the minds of the laity which, in turn, has led to an over-abundance of illnesses and scourges in Westernized populations.[16]

Today's common diseases were relatively rare in the 1930s, but are ever-increasing among so called "well-fed" populations. However, research even shows that ancient peoples were also cursed with these diseases that came from heavy meat consumption. In the book of Exodus, God addressed the Israelites thus:

> If you diligently heed the voice of the Lord your God and do what is right in His sight, give ear to His commandments and keep all His statutes, I will put none of the diseases on you which I have brought on the Egyptians.[17]

What types of diseases were inflicted on the Egyptians? Dr. Marc Armand Ruffer, a Pale pathologist has, along with his associates, performed over 36,000 autopsies on Egyptian mummified remains of Pharaonic royals. Ruffer's research demonstrates that most of the diseases striking the Egyptian royalty bear an uncanny resemblance to those diseases killing us today: various forms of heart disease (atherosclerosis), osteoporosis, obesity, stroke, tooth decay, arthritis, diverticulitis of the colon, early sexual development in children, and cancer. These diseases were being caused mainly by excessive meat-eating and wayward diets. This could be concluded by examining the presumably well-fed royal babies with cholesterol deposits and narrowing of the artery lumens.[18]

Animal-based diets are also linked to breast cancer since high oestrogens are a predominant factor in breast cancer. Meat-eating women have higher levels of oestrogen in the urine than vegetarian women, according to research.[19]

Humans are herbivores with anatomical and physiological features, but in many of today's societies they have been taught to eat both animal flesh and plants. Thus, we humans are omnivores by culture only. And though our meat-eating culture has developed a number of myths (historical and 'medical') to bolster the claim that flesh-eating is 'natural' or that we have 'evolved' to eat meat, the facts of anatomy clearly indicate otherwise (William C. Roberts, 2009).

Humans must discontinue consuming animal products, otherwise we will definitely see a higher trend in many forms of degenerative diseases, especially heart diseases and cancers in those nations that consume these foods.

According to William C. Roberts, if we "look at various characteristics of carnivores (meat eaters) versus herbivores (non-meat eaters), it doesn't take a genius to see where humans compare".[20]

According to biologists and anthropologists who study our anatomy and our evolutionary history, humans are herbivores that are NOT well suited to eating meat. Unlike natural carnivores, we are physically and psychologically unable to rip animal's limbs from limb and eat and digest their raw flesh. Even cooked meat is likely to cause health problems in human beings, but not so in natural carnivores, that DO NOT suffer from food poisoning, heart disease, and other ailments.[21]

Milton Mills, M.D. stated that:

> "Omnivore" doesn't mean 50% plants and 50% animals. Many of my critics consider chimpanzees to be omnivores, but 95-99% of the chimp diet is plants, and most of the remainder isn't meat, it's termites. If humans are omnivores, then the anatomical evidence suggests that we're the same kind: the kind that eats almost exclusively plant foods.

Primates may have adapted and evolved while consuming small amounts of animal flesh, but this does not mean that eating animal flesh was healthy for our ancestors. Today, we know that animal foods can bring about undesired consequences to human health. Even eating minute amounts of animal flesh has been proven to cause disease in humans, especially heart disease. If eating meat were so natural, then why do animal products cause so much damage to human health? It surely makes more sense to eat more plant vegetation.

Doctors must focus more on diet and nutrition to render diseases preventable. Thomas Edison once said: "The doctor of the future will give no medicine, but will instruct his patient in the care of the human frame, in diet, and in the cause and prevention of disease."

Plant vegetation *is* the main food humans should be consuming. Our physiological anatomy and appearance, according to scientific literature on this subject, prove that Homo sapiens has transformed towards the naturally herbivorous species in the evolutionary tree of primeval life.

Humans should not be eating both plants *and* animal foods to sustain optimal health. Human diseases have already been documented and scientifically proven to be on the increase in today's western societies and among human populations that consume the most animal-based foods. As W.S. Collens asserts: "Atherosclerosis affects only herbivores that eat animal produce. Dogs, cats, tigers, and lions can be saturated with fat and cholesterol, and atherosclerotic plaques *do not* develop."[22]

This is corroborated by W.C. Roberts who wrote: "The only way to produce atherosclerosis in a carnivore is to take out the thyroid gland: then, for some reason, saturated fat and cholesterol have the same effect as in herbivores."[23]

The Heart to Primary Disease Prevention: a "Whole, Rich, Plant-Based Diet"

The only substantive healthy foods for humans are plants and their derivatives, which nature has provided spontaneously and which have prevailed in the evolutionary process: in the beginning when plant vegetation and seeds matured, blossomed and sprouted into wonderful, healthy, edible fruits and other plant foods.

According to medical explorers, who study cultures from different regions around the world, populations that live isolated from the industrial westernized world live longer and more satisfying lives. Why? This is because they are less inflicted by industrial toxins and contaminated water, and they eat minimum amounts of processed/refined and animal foods. Their diets are based on natural plant-based foods and only small amounts of meat and dairy produce. Some also eat the pith and seeds of fruits that have anti-cancerous properties.

We need to adapt and pursue the virtuous examples of the many, small, isolated civilizations that live in specific regions around the world and who live longer and more sustainable lives. Recognition of these cultural-studies have been done, and they have all proved that whenever we eat meat and any form of animal produce and processed industrialized foods our health suffers tremendously.

Unhealthy food choices *may well be* the main factor that *contributes to many forms of acute and chronic-related diseases*.

Ty Bollinger shrewdly quoted that we all need to *"step outside the box"*[24] and truly esteem the natural beauty of the world, where modern man has gone wrong in his eating habits. Nature has provided man with foods grown from its soil: nutritious, dense, vibrant and most colorful-protective plant foods. Let us take advantage of these foods, as it is these foods that can keep us alive for as long as *Nature* purposefully intended.

Notes for chapter 1

1. Dyson, Freeman (1999), "Origins of Life" (Second Edition), Cambridge University Press.
2. Astrobiology Web. (2014). Wellcome Trust: Metabolism May Have Started in Early Oceans Before the Origin of Life". Retrieved from http://en.wikipedia.org/wiki/Abiogenesis. Keller, Markus A.; Turchyn, Alexandra V.; Ralser, Markus (25 March 2014). "Non-enzymatic glycolysis and pentose phosphate pathway-like reactions in a plausible Archean ocean". *Molecular Systems Biology*, **10** (725). doi:10.1002/msb.20145228. Retrieved from http://en.wikipedia.org/wiki/Abiogenesis
3. Sagan, Carl. (1973). *The Cosmic Connection*, Anchor Press, Doubleday Garden City. New York.
4. Human Evolution. Retreived from http://www.encyclopedia.com/topic/human_evolution.aspx
5. Milton, K. (1999). Nutritional Characteristics of Wild Primate Foods: Do the diets of our closest living relatives have lessons for us? *Nutrition*, **15** (6), pp. 488-498.
6. Ibid.
7. Ibid.
8. See, for example: Danby, F.W. (2009). Acne, Dairy and Cancer. *Dermato-Endocrinology*, **1**, pp 12-16; Melnik, B. (2009). *Milk Consumption: aggravating factor for acne and promoter of chronic diseases of Western societies*. JDDG, 7, pp 364-370; Ornish, D. (1990). Lifestyle Changes Reverse Coronary Heart Disease. Medical Science: The Lifestyle Health Trial. *Lancet*, **336**, pp. 129-133; Elio Riboli.(2001). The Role of Nutrition in Preventing and Treating Breast and Prostate Cancer: The European Prospective Investigation into Cancer and Nutrition (EPIC): Plans and Progress. *The Journal of Nutrition*, **131**, pp. 170S-175S.
9. William Roberts (1991). Editor in chief of the *American Journal of Cardiology*.
10. Roberts, C.W. (2010). Humans are Herbivores. Retrieved from https://www.youtube.com/watch?v=yUa814suU9A
11. Ibid.
12. A service of E/The Environmental Magazine (2002) Retrieved from http://www.rense.com/general20/meant.htm; The Ecologist. (1970-2000). The Myths of Vegetarianism. Retrieved from www.theecologist.info/page14.html.
13. Ibid.
14. W.C. Roberts, ibid.
15. Blandche, L. (1979). Cancer and Other Diseases from Meat Consumption: Leaves of Healing. Santa Monica, CA

16. Phillip day. (2001). Health Wars: White Lies and Porky Pies. The unsettling reasons why Meat and Dairy can kill. pp.145-146. Credence Publications.
17. Ex 15:26
18. Loma, L. (1984). Mysteries of the Mummies: Slide-tape program produced by Loma Linda at the University School of Health.
19. Schultz, T. *et al.* (1983). Nutrient intake and hormonal status of premenopausal vegetarian Seventh Day Adventist and premenopausal non-vegetarians. *Nutrition and Cancer,* **4**, pp. 247-259.
20. Roberts, W.C, M.D. Physiological Features and Anatomical proof of Evolutional Herbivore Origins of Homo-Sapiens (Modern Man).
21. Milton Mills. *The Comparative Anatomy of Eating.* http://www.adaptt.org/Mills%20The%20Comparative%20Anatomy%20of%20Eating1.pdf
22. Collens, W.S. (1969). Atherosclerotic disease: an anthropologic theory. *Medical Counterpoint,* **1,** pp.53–57.
23. Roberts, W.C. (1990). We think we are one: we act as if we are one, but we are not one. *American Journal of Cardiology,* **66,** 896.
24. Bollinger, Ty. (2006). *Cancer Step outside the Box.* Published by Infinity 510^2 Partners.

2

Cancer: Causes and Prevention

One must not forget that recovery is brought about not by the physician, but by the sick man himself, by his own power, exactly as he walks by means of his own power, or eats, or thinks, breathes or sleeps.

Georg Groddeck
The Book of the It (1923)

Cancer is a result of normal human cells losing their ability to function adequately and dividing uncontrollably in a part of the body. Cancer needs the right environment to grow, multiply and spread to other organs of the body. If Homo sapiens' *internal homeostatic cellular environment* is not controlled, or a carcinogen/agent is not removed, human DNA damage develops and increases, and body cells may change abnormally into microscopic, benign or malignant, tumours: that can multiply in a disordered way, grow, and finally metastasize (spread) to other parts of the body.

Lack of essential nutrients and the build-up of toxicity within the body (toxic-overload), can only worsen the problem. In incidences like these, free radicals within the body can accumulate, and can alter or damage our genetic-blueprint "Deoxyribonucleic Acid" (DNA), leading to gene alterations and oxidative stress, and affecting the aging process, which itself increases the risk of disease.[1]

A Warning Sign ... and a Friend!

Paradoxically, cancer could be considered to be our friend. In most, but not all cases, if you develop cancer, then it is a scientific fact that the cause could be your lifestyle and bad long-term dietary eating patterns. It is a signal from the body telling you that you are not looking after it well enough.

I personally believe that cancer is not a disease, but a warning sign. It is the body's way of telling us that something is out of balance and needs to be corrected. Our body is a marvelous machine that needs to be well-nurtured and maintained, with high-quality and healthy nutritious foods combined with a healthy lifestyle. The human body is showing us the "symptoms of disease", and it will not tolerate any ghastly eating habits nor lackadaisical lifestyles.

Cancer is actually a very slow-growing process. Most often, tumours can take up to 2-20 years to develop, becoming large enough to be recognized and detected by using a mammogram (in the case of breast cancer) or by other medical diagnostic equipment. Many years of breathing, digesting, absorbing and exposing the human body to environmental obnoxious pollutants, unnatural foods and chemicals will eventually take their toll, and, ultimately, can lead to the development of chronic disease.[2]

Everyday factors can increase our risk of developing cancer because they eventually produce free radicals (oxygen-reactive species) within our bodies. Many scientists now believe that this is one of the major risk factors that contribute to the aging process and towards the development of serious illnesses. This is why antioxidants, found most abundantly in plant foods, help bind to these oxygen-reactive species and which can, ultimately, reduce the damage done to human body cells and tissues: especially to our cell membranes, mitochondria's and our double-stranded DNA.[3]

Factors that can contribute to cancer

- Sunlight's Ultraviolet rays
- Cosmos Rays & X-rays
- Ionizing Radiation
- Nuclear Radiation
- Pesticides/Herbicide residues
- Tobacco and smoking
- Hormonal therapies
- Chemotherapeutic Drugs
- Medications
- Cosmetics
- Diet
- Nutritional deficiencies
- Parasites
- Viruses
- Bacterial Infections
- Fungal Infections
- Cellular Oxygen Deficiency
- Oncogenes
- Asbestos
- Environmental pollutants
- Excess Alcohol consumption
- Genetic predisposition
- Blocked Detoxification Pathways
- Ingesting Food Additives
- Toxic Metal Syndrome
- Dental Amalgam Fillings/Root Canals
- Use of Street Drugs/other Drugs
- Chronic Physical or Mental Stress

Cancer: There is a Limit to How Much the Human Body can Handle

What causes cancer? There are many contributing factors to the development of cancer. Nevertheless, with a healthy immune system, the body is usually able to identify the existence of cancer and wipe out slow-forming microscopic tumours growing within the body. It has now been shown that each one of us may actually get cancer several times in our lifetime. The amazing ability of the human immune system can most certainly build protective walls and help eliminate small, benign and malignant tumours. However, if the body's immune system has been immunosupressed, it cannot carry out its normal defensive functions, and cancer may perhaps someday win the battle – unless of course, the underlying causes or risk factors are recognized and removed.[4]

Let us take an example: A **carcinogenic agent** is the "**seed**" that can **initiate** cancer development, and **diet** is the "**fertilizer**" that **promotes** the growth of cancer. If the cancer continues to be fertilized, then it can grow and spread (metastasize) to other organs of the body. For example, the carcinogens in cigarette smoke are the "seeds" for initiating cancer, whilst animal products are the "promoters" for cancer growth.

If we do not fertilize and feed the cancer, the seeds may *not* sprout.

The goal of a good anti-cancer diet is to get as much micronutrients per calorie into the body as possible. An anti-cancer diet, in other words, should consist of a "Rich Micronutrient/Phytonutrient-Dense Plant-Based Diet".[5]

Focus on Vegetables and Fruits

An assortment of a specific anti-cancer diet consisting mostly of fruits and vegetables have many healthy, healing and appealing colors: dark green, red, blue, yellow, purple and orange. A colored variety and a synergistic effect of all types of plant foods is fundamental to cancer prevention. Herbs and spices, nuts, seeds and legumes are also fundamental plant-food sources, because they also contain healthy nutrients: vitamins and minerals, fibre, antioxidants and the essential omega-3 and omega-6 fatty acids.

The consumption of plant foods that contain powerful and protective synergistic chemicals (phytonutrients and antioxidants) has been scientifically proven to protect us from cellular damage and the build-up of internal/external environmental pollutants, as well as to help reduce the amount of oxygen-reactive species (free radicals) that react within the cells of our body, that may lead to oxidative stress. These plant components are extremely beneficial, are contained in nutritionally color-dense plant foods and are critical for the preservation to human health.

According to M. Anderson, "A plant-based diet and bettering lifestyle changes can ultimately reduce our risk of developing any form of disease, including cancer."[6]

The Importance of Nutrition for Cancer Patients

Why is nutrition so important? It is because most cancer patients do not die from their tumours, but they die from:

1. Malnutrition
2. Toxaemia
3. Opportunistic Infections, and
4. Despair.

The Centre for Advancement in Cancer Education believes that 90 percent of all cancers can be eliminated **through environment and lifestyle choices alone**. Proper nutrition can address the first three factors mentioned above, whilst stress management and patient support can address the fourth (despair).[7]

According to Dr. Susan Silberstein, who strongly states that:

> We need to be dealing with cancer in our kitchens. Tens of thousands of research articles have been published in hundreds of biomedical journals worldwide, documenting the scientific relationship between diet and cancer survival. That is not just prevention but, actually, survival. This is corroborated by Dr. Susan Silberstein's vast experience in counselling thousands of cancer patients. She, as well as many other

clinical medical scientists, has personally seen many fourth-stages, advanced, terminal cancer patients who were given only months, sometimes weeks or even a few days to live, and who have fully recovered by following dietary programmes.[8]

A Healthy Advice: The Late Dr. David Serven-Schreiber: A New Way of Life

The first thing he learned was that everyone of us has cancer cells, yet, only 1 out of 3 in the Western world will die of cancer; while 2 out of 3 will not. That means for those 2 out of 3 fortunate persons must have natural defenses that can slow-down the disease.

Food is something you do 3 times a day to your body every day. And we now know that phytochemicals that are part of everything you eat has a profound influence on your physiology. What you eat has a big influence on tumour development. The beauty of food is that it doesn't just target one mechanism of cancer – it plays on a whole range of the keyboard.

E.g., Turmeric will act on surly aspects of cancer biology; cabbage or broccoli contains sulphorphane and Inole-3-Carbinols will play another aspect; green tea contains EGCG, polyphenols also play on another aspect. Then you can add raspberries or blueberries, which indeed contain, ellagic acid, anthocyanidins and proanthocyanidins, in bringing in another range of phytochemicals, which will yet act on a different mechanism on cancer. We are building an anti-cancer biology. Anybody can do that and we just need to educate people and tell them how they can do it by themselves.[9]

The World Research Cancer Fund says that "the goal should be for people to eat no more than eleven ounces of red meat per week". In the U.S they are consuming around eleven ounces of red meat per day. Red meat is far more expensive than lentils, beans, and cereals or soy and tofu. These are good sources of vegetable protein that can replace meat and act as anti-cancer agents, which will allow us to reduce considerably our intake of animal products, especially red meat.[10] We now know that any form of animal flesh increases cancer risk.[11]

Similar, the vegetables that are best for you are actually some of the cheapest to buy. You can add garlic, purple cabbage and cauliflower, kale, beets, Brussels sprouts, flaxseeds, broccoli, onions and leeks to almost every dish you cook: if you do that it does not cost a lot of money and it builds a strong anti-cancer diet.[12]

If we do not tell the general public, that science clearly shows how much people can protect themselves from their own disease in making their own choices; that when they start to take the lives in their own hands they can not only prevent disease, but actually reverse it. If we don't tell them that or don't give them the chose in doing so, we have given them false hopefulness. We know that false hopefulness can actually promote the growth of cancer.[13]

Let us study the future of health care and look forward to a new and healthier approach, that focuses mainly on nutrition and lifestyle medicine for disease prevention and disease reversal.[14]

Notes for chapter 2

1. Doll, R. et al. (1981). The causes of cancer: quantitative estimates of avoidable risks of cancer in the United States today. *Journal of the National Cancer Institute*, **66**, pp.1191-308. Willett W. (1995). Diet, nutrition, and avoidable cancer. *Environ Health Perspective*, **103**, pp.165-170; Ames, B, et al. (1993). Oxidants, antioxidants, and the degenerative diseases of aging. *Proceedings of the National Academy of Sciences*, **90**, pp.7915-7922.

2. Ibid.

3. Ames, B. et al. ibid.; Doll, R. et al. ibid.; Heber, D. & Bowerman, S.(2001). *What Colour Is Your Diet?* Published by Harper-Collins/Regan Books. New York.

4. Block, G., Patterson, B., & Subar, A. (1992). Fruit, vegetables, and cancer prevention. A review of the epidemiological evidence. *Nutrition and Cancer*, **18**, pp.1-29; World Cancer Research Fund.(1997). Food, nutrition and the prevention of cancer: a global perspective. *Washington, DC; WCRF and American Institute for Cancer Research*; Steinmetz, K.A. & Potter, J.D. (1991). Vegetables, fruits, and cancer. *Epidemiology. Cancer Causes Control*, **2**, pp.325-337; Liu, R.H. et al. (2003). Health benefits of fruits and vegetables are from additive and synergistic combinations of phytochemicals. *American Journal of Clinical Nutrition*, **78**(3), pp. 517S-520S.

5. Gao, C.M., Tajima, K., & Kuroishi, T. et al. (1993). Protective effects of raw vegetables and fruits against lung cancer among smokers and ex-smokers: a case-control study in the Tokai area of Japan. *Japan J. Cancer Research*, **84**(6): pp.594-600.

6. Anderson, M. (2009). *Healing Cancer from Inside Out* (1st ed), Part 2: Reversing Cancer: Rebuilding the Immune System, (p.57). Published by www.RaveDiet.com

7. Silberstein, S. (2012). *Cancer Survival: 3 reasons why most people do not die from their cancer*. Centre for Advancement in Cancer Education. Retreived from http://www.youtube.com/watch?v=AtnEPAG5_H8

8. Silberstein, S. *Cancer Survival*: Retreived from www.beatcancer.org

9. Schreiber, S.D. (2008). Anticancer – a new way of life: An interview with David Servan-Schreiber. Retreived from https://www.youtube.com/watch?v=2IwiQm5QaTs

10. Ibid.

11. World Cancer Research Fund. (2007). Food, nutrition, physical activity, and the prevention of cancer: A global perspective. *American Institute of Cancer Research*. Washington, DC. Barnard ND, Nicholson A, Howard JL. (1995). The medical costs attributable to meat consumption. *Prev Med*, **24**. pp.646-655. Chang-Claude J, Frentzel-Beyme R, Eilber U. (1992). Mortality patterns of German vegetarians after 11 years of follow-up. *Epidemiology*, **3**, pp.395-401. Chang-Claude J, Frentzel-Beyme R. (1993).

Dietary and lifestyle determinants of mortality among German vegetarians. *Int J Epidemiol*, **22.** pp.228-236. Schreiber "Anticancer", 2008.

12   Schreiber "Anticancer", 2008.

13   Ibid.

14   Phillips RL. (1975). Role of lifestyle and dietary habits in risk of cancer among Seventh-day Adventists. *Cancer Res*, **35**(suppl), pp.3513-3522.

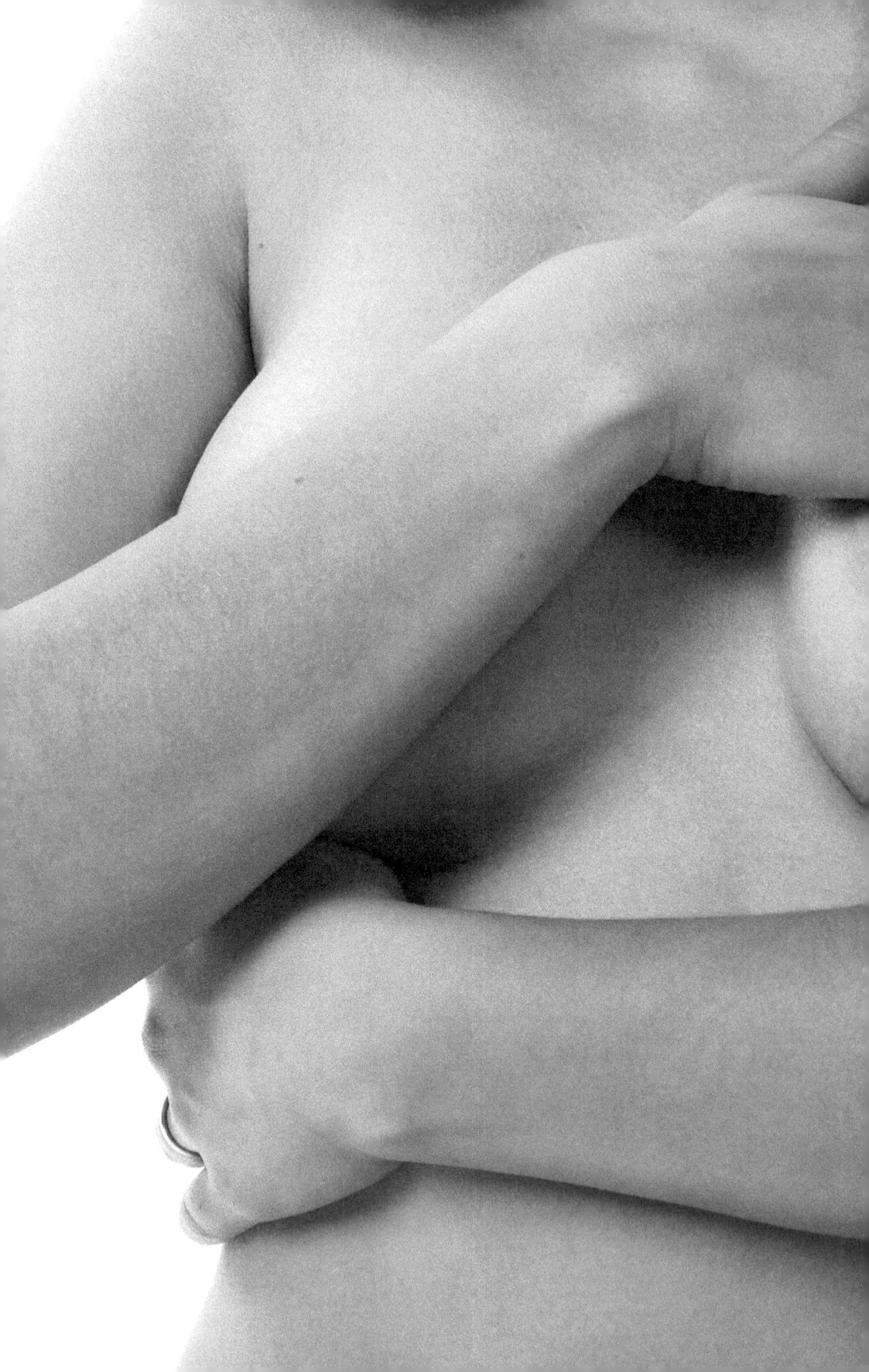

3

Breast Cancer

Cancer is a journey, but you walk the road alone. There are many places to stop along the way and get nourishment – you just have to be willing to take it.

Emily Hollenberg
Breast cancer survivor

Breast cancer is due mainly to the abnormal proliferation and development of normal human cells in the breast. These cells can then transform into benign (fibrosis/or cysts) or malignant (*in situ* or invasive) tumours. Not all tumours are cancerous.

Benign tumours can cause problems – they can grow large enough and press on healthy organs and tissues, but they are rarely invasive.

Malignant tumours are a group of cells that can grow into (invade) surrounding tissues or spread (metastasize) to other distant areas around the body: via the blood or the lymphatic system (vessels and nodes), and cause harm to the body.

Breast cancer is most commonly found among women, but a small minority of men can also develop the disease. To fully comprehend the biological mechanisms involved in the development of breast cancer, we must fully comprehend some basic knowledge about and the morphological physical structures of the breast in women.

There are many lobules that make up the female breast (milk-producing glands), the ducts (small tubes that carry the milk from the nodules to the nipple), and the stoma (fatty tissue and connective tissue surrounded by ducts and nodules, blood and lymphatic vessels).

There are different types of breast cancers. They are usually named after the types of cells from which the cancer is thought to have originated.

There are non-invasive ductal or lobular carcinomas, where the tumour cells have not yet spread to the surrounding tissue, or outside the ducts or lobules. These breast tumours are usually called *in situ*, because they have not yet spread outside or beyond the region of these parts of the breast.

If malignant cells spread beyond the lining of the ducts or lobules invading into the surrounding breast tissue and travel to the lymphatic system, this may be considered as an invasive breast cancer. Breast malignant cells have the ability to metastasize (spread) to other organ tissues. In breast cancer, the main sites of metastasis are the bone, the brain, the liver and the lungs, but it is still considered as a breast cancer as the primary malignant tumour originated from the breast.

Most breast cancers start in the cells lining the ducts (the channels in the breast that carry milk to the nipple) and spread into the surrounding breast tissue. This is known as an invasive ductal breast cancer. This is the most common type of invasive breast cancers. In fact, between 70-80 percent of all known breast cancers are of this type. The other 10-20 percent represents invasive lobular breast cancers and finally rarer types of inflammatory breast cancers account for the other 1-3 percent. There are other, less common, types of breast cancers (sarcomas) that start in connective tissues: such as the muscle tissue, the fat tissue and the blood vessels. Sarcomas of the breast are very rare.

In the late 1970's, breast cancer struck one in every eleven women. In the 1980's the rate went up to one in ten. By 1992, the rate was one in eight. In the United States, the annual numbers of new cases went up from 73,000 to 135,000.[1]

Breast Cancer is Preventable

The link between diet and cancer is not new. An article in *Scientific American*, as early as 1892, reproduced an observation stating that "cancer is most frequent among those branches of the human race where carnivorous habits prevail".[2] The National Research Council published a report, *Diet, Nutrition, and Cancer*, that showed voluminous evidence that linked specific dietary factors to cancer of the breast and other organs.[3]

Unfortunately, such vital information remained unused and sat collecting dust at cancer research centers. There has never been an organized effort to give women the opportunity nor the information needed, to make their own decisions about cancer prevention.

Numerous research studies have shown that cancer is much more common in populations consuming diets rich in fatty foods: Meat and dairy produce, whilst cancer prevalence is much less common in countries eating diets rich in grains, vegetables and fruits. One reason is that foods affect the action of hormones in the body. They also affect the strength of the immune system and other human aliments. Likewise, whole fruits and vegetables contain a variety of vitamins, minerals, antioxidants and other phytonutrients that help protect the body from disease. In contrast, recent research shows that animal products contain potentially carcinogenic compounds which may contribute to increased cancer risk.[4]

In addition to tobacco use and diet, other factors, including physical activity, reproductive and sexual behavior, bacterial and viral infections, and exposure to radiation and chemicals, may also contribute to the risk of certain forms of cancer.[5]

Estimated Percentages of Cancer Due to Selected Factors [5,6]

Diet	35% to 60%
Tobacco	30%
Air and Water Pollution	5%
Alcohol	3%
Radiation	3%
Medications	2%

Lower Rates of Breast Cancer in Asian Countries

Why is it that Japanese and Asian women have a much lower mortality rate of breast cancer? Japanese women even have one of the lowest rates of developing breast cancer. What is the reason for this? Medical scientists first realized that there could be a link between cancer and diet when they realized that Asian populations, when compared to Western Europeans and Americans, had the lowest rates of cancer. These differences are not due to genetics, nor due to something in

> **FACT BOX**
>
> Most people consider so called "white meat" as healthy. **It is not!**
>
> Fish, chicken or any so-called "white meat" contain high levels of cholesterol, saturated fat, and industrial pollutants.

the soil, air or water. Many scientific studies have concluded that one important factor is the food we eat.[6]

Many Asian nations eat and drink huge amounts of whole-grain rice (black, red, white or brown) including mushrooms, soy produce, green tea, seaweeds and large amounts of vegetables and fruits. Asian populations consume no, or very small quantities of, dairy products and very little meat, with the exception fish. Fish is a sizeable part of their animal diet which indeed contains a significant amount of omega-3 fatty acids, which has been known to be beneficial for the protection of human health.

Fish does indeed contain adequate amounts of essential fatty acids, but it is healthier for you to obtain your Omega-3 fatty acids from plant sources, that are highly nutritional and are synergistic types of foods. Plant foods like Green-leafy vegetables, flax seeds, hemp seeds and walnuts, are better food alternatives in retaining these essential fats: via fish and omega-3 supplementations.[8]

According to Dr. John McDougall: The Great Nutrition Debate

If I were to ask you to design a diet that would keep people thin for a lifetime, what populations would you pick? China, Korea, Japan, Indonesian and African countries will probably come to mind. It is obvious. This is not surprising. These populations normally live on a starch-based diet with the addition of fruits and vegetables, whilst rich foods are kept to a minimum. Nevertheless, when these people move to the United States or change their diet in their own land;

they consistently get fat, develop heart disease, rheumatoid arthritis, diabetes, cancer and other diseases.[9]

The Cancer Migration Paradox

Unfortunately, however, as the traditional diets of Far East Asia have been abandoned and westernized diets have taken over, breast cancer rates have escalated. Clinical studies have linked these higher cancer incidence rates to the quadrupling of the consumption of animal-based food products.

For example, when Japanese women emigrate to the U.S.A., their breast cancer risks go up due to a change in their lifestyle and westernized eating habits. Evidently, cancer risk then may have more to do with one's lifestyle, diet or the environment than with any genetic factor. The most obvious change is *diet*.

A Japanese woman is less likely to develop breast cancer than an American woman, and even if she does get cancer, she is much more likely to survive. Why?

The first theory suggested that Japanese women are mostly thin and not overweight. This is important because we now know that body fat actually acts as a storehouse for producing oestrogens (female sex hormones) that act as fertilizers for breast cancer growth, making the disease more aggressive.

But that is only part of the story. Diet plays a role, even if the female is not overweight. If a woman is on a diet that is high in fat and very low in fibre; such diets also tend in increasing the amount of oestrogens in her blood. Excess male and female sex hormones are filtered in the liver, are transported to the bile duct, then into the intestinal track, and finally removed via the large colon. The only problem is that this excretion system depends on one important plant component: it depends on plant roughage, or "plant fibre".[10]

In a 2007 publication, J.E. Cade concluded as follows:

Excess oestrogens or other hormones need to bind to fibre in order for them to be removed and excreted from the body as waste. Where do we get fibre from? "Plants." Animals have muscles and bones for stability and support, whilst plants have fibre. Also, phytoestrogens found in natural plant foods (flax seeds, sesame seeds, soya beans and other legumes, including nuts) contain natural phytoestrogens. These very weak plant estrogens can occupy the oestrogen receptors on breast cells, displacing normal estrogens. The result is less oestrogen stimulation of each cell. Are there any phytosterols or plant roughage (fibre) in animal products? No! There is not.[11]

In past decades, there has been major health debates about soya products: some clinical scientists say soya is healthy, while others disagree. Who is right?

A recent analysis has concluded that soya isoflavones' oestrogen-like effects are probably too weak to have any significant beneficial effect on breast cancer tissue in healthy women, even breast cancer survivors.[12] Two ounces a day, by the way, is the typical amount of soya included in traditional Chinese and Japanese diets.[13] Breast and prostate cancer rates are four to six times lower in Japan and China than in Western countries, and laboratory studies have shown that isoflavone from soya can inhibit the growth of both breast and prostate cancer tissues.[14]

One should remember that soya is a large part of the Asian diet, and apparently it may or may not affect the welfare or the health of

---

FACT BOX

It is important to note that 40 grams of soya protein can increase Insulin-like Growth factor 1 (IGF-1), so try to eliminate soy products from your diet, including milk. A study done on milk and soya protein proved that these two food choices increased IGF-1 by 36% (milk) and 69% (soya). *J. Clin, Endrocrinol Metab*, **March 2003, 88(3), pp.48-54.**

populations that consume high amounts of soya produce. But, soya may also offer protection because Asian populations consume vast amounts of fruits and vegetables with soya. Soya may even be a good alternative to cow's milk. But, it is reiterated, and so, more clinical studies on soya are needed to confirm these conclusions.

My opinion is that for the time being, and until more scientific evidence proves that soya is beneficial to human health, we should reduce or avoid consuming any type of food made from soya. After all, there are a variety of other foods to choose from a plant-based diet that are highly healthy and nutritious, without having to consume or worry about the health consequences contributing to soya products.

More on Dietary Fibre in Decreasing Oestrogen Levels

Fibre is extremely essential to human health, due to it's many biochemical health functions: increases stool bulk, helps lower blood sugar levels, weight loss management and controls sugary surges, binds and helps eliminate excess toxins, hormones and cholesterol, lowers the formation and risk of Hemorrhoids, Diverticulitis, Irritable Bowel Syndrome, Colitis and Crohn's disease. Dietary fibre also aids

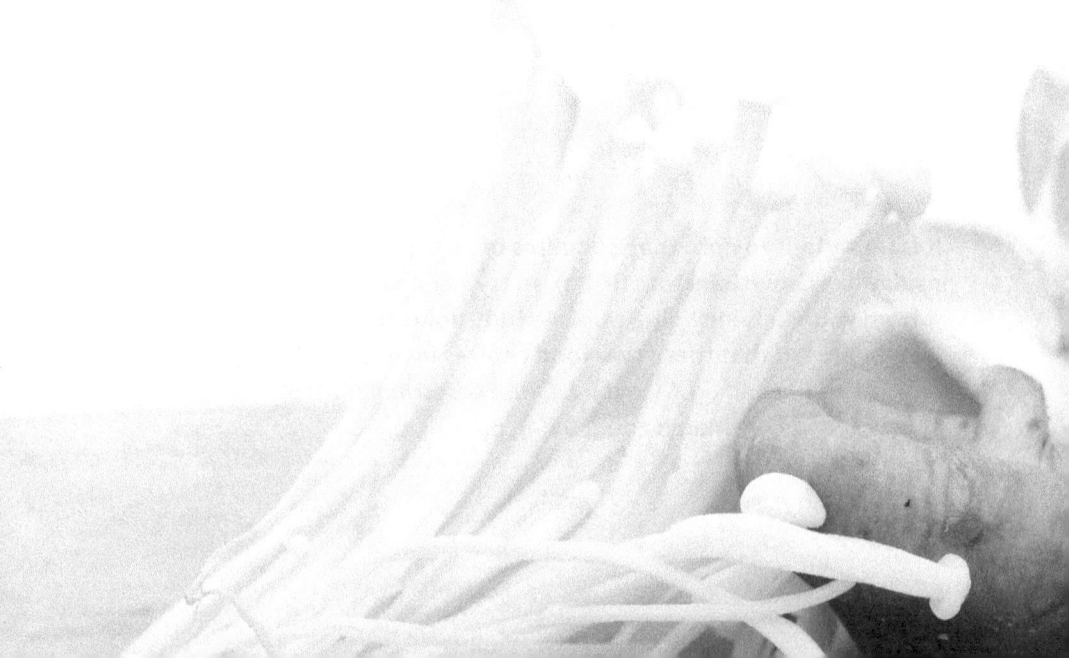

towards the function and metabolism of foods in the GIT by our good bacteria flora.

Also, if there is no plant roughage within the intestinal track, oestrogens are reabsorbed back into the body's blood circulation – a process called *enterohepatic recycling* (intestinal/liver circulation). Evidently, one important aspect of fibre is that it can lower oestrogen levels in the body.

Animal fat is another health concern. Although scientists continue to debate the role of fat in cancer, substantial evidence shows that the more fat, especially saturated/trans-fatty acids, there are in the diet, the greater the risk of breast cancer.[15] Plant-based foods contain denser amounts of nutrients, fibre, vitamins, high antioxidant contents and other essential micronutrients like selenium, calcium, iron, potassium, magnesium and zinc, etc... to help protect the body. As we look at these essential nutrient components, it is relevant to remember that each one, unaccompanied, is just part of the problem. It is not animal fat alone that invigorates cancer nor is it just fruits and vegetables that entirely prevents cancer, but rather *the combined effect of the nutrients and poisons we put on our plate three times per day.*

Many Western societies tend to eat a diet in which a high percentage of calories come from processed/refined foods, oils, and other foods of animal origin, and smaller amounts of plant foods. In later chapters we shall discuss in more detail the main causes and strategies needed for the best prevention and management treatments of breast cancer and other chronic diseases. We shall see that there is a link between diet and cancer and and the correlation to what types of foods: Plants vs. Animal. We shall also see that both the "quantity" *and the* "quality" of fats, oils, protein, including carbohydrates within the nutritional components of foods will most certainly make a tremendous amount of difference between health prevention, control and patient survival or to the development of disease.[16]

Notes for chapter 3

1   U.S. General Accounting Office. (1991). *Breast Cancer: Prevention, Treatment and Research.* GAO/PEMD-92-12.

2   Commission on Life Sciences & Nutrition, and Cancer Committee on Diet. (Januaty 15, 1982). *Diet, Nutrition, and Cancer.* National Academy Press.

3   Ibid.

4   Skog KI, Johansson MAE, Jagerstad MI. (1998). Carcinogenic heterocyclic amines in model systems and cooked foods: a review on formation, occurrence, and intake. *Food and Chem Toxicol,* **36**, pp. 879-896. Physicians Committee for Responsible Medicine. Food for Life, Cancer Project: Factors Contributing to Cancer. Retreived from http://www.pcrm.org/health/cancer-resources/diet-cancer/facts/factors-contributing-to-cancer

5   Minamoto T, Mai M, Ronai Z. (1999). Environmental factors as regulators and effectors of multistep carcinogenesis. *Carcinogenesis,* **20**(4), pp.519-527. Cummings JH, Bingham SA. (1998). Diet and the prevention of cancer. *BMJ,* 317, pp.1636-1640. Ibid.

6   Armstrong, B., Doll, R., *et al.* (1975). Environmental factors and cancer incidence and mortality in different countries, with special reference to dietary practices. *International*

Journal of Cancer, **15**, pp.617-631. Hirayama, T., *et al.* (1978). Epidemiology of breast cancer with special reference to the role of diet. *Preventive Medicine*, **7**, pp.173-195. Hirose, K., *et al.* (2007).Dietary Patterns and the Risk of Breast cancer in Japanese Woman. *Cancer Science*, **89**(9), pp.1431-1438.

7. Hirose, K., *et al.* (2007). Dietary Patterns and the Risk of Breast cancer in Japanese Women. *Cancer Science*, **89**(9). pp.1431-1438.

8. Hirose, "Dietary Patterns", 2007.

9. Sponsored by the USDA. (February 24th, 2000). *The Great Nutrition Debate*. The debate features speakers such as Dr. John McDougall, Dr. Robert Atkins, Dr. Barry Sears, Dr. Denise E. Bruner, Dr. Dean Ornish, Dr. Morrison, C. Bethea, Dr. Keith T. Ayoob, and a small presentation by Dr. Eric Westman. Retreived from http://www.youtube.com/watch?v=J29eNPohRyk

10. Cade, J.E., *et al.* (2007). UK Women's Cohort Study Steering Group. Dietary Fibre and Risk of Breast Cancer in the UK Cohort Study. *International Journal of Epidemiology*, **36** (2): pp.431-438.

11. Ibid.

12. Messina, M.J. *et al.* (2008). Soy isoflavones, estrogen therapy, and breast cancer risk: analysis and commentary: *Nutrition Journal*, **7**(17). doi: 10.1186/1475-2891-7-17.

13. Nagata, C. *et al.* (1998). Decreased serum total cholesterol concentration is associated with high intake of soy products in Japanese men and women. *Journal Nutr*, **128**(2), pp.209-13.

14. Adlercreutz, H. (2002). Phyto-oestrogens and cancer. *Lancet Oncol*, **3**(6), pp.364-373.

15. Chlebowski, R.T., *et al.* (2006). Dietary fat reduction and Breast cancer outcome: Interim efficacy results from the Women's Intervention Nutrition Study. *Journal of the National Cancer Institute*, **98**(24): pp.1767-1776. Carroll, K.K., Braden, L.M. (1985). Dietary fat and mammary carcinogenesis. *Nutrition and Cancer*, **6**(2), pp.54-59. Rose, D.P., Boyar, A.P., Wynder, E.L. (1986). International comparisons of mortality rates for cancer of the breast, ovary, prostate, and colon, and per capita food consumption. *Cancer*, **58**(23), pp.63-71.

16. Skog, K., Johansson, M.A.E, Jagerstad, M.I. (1998). Carcinogenic heterocyclic amines in model systems and cooked foods: a new review on formation, occurrence, and intake. *Food and Chem Toxicol*, **136**, pp. 879-96. Cummings J.H., Bingham S.A. (1998). Diet and the prevention of cancer. *BMJ*, **317**, pp.1636-40. Chavez, A. *et al.* (1998). Diet that prevents cancer: recommendations from the American Institute for Cancer Research. *Int. Journal Cancer Suppl*, **11**, pp.85-89.

4

Making the Right Food Choices

*Let food be thy medicine,
and medicine be thy food.*

Hippocrates[1]

Select the Best and Healthiest Foods

According to Dr. Joel Fuhrman,

> Whenever we try to get nutritious excellence from foods that have synthetically-added components; that is food whose naturally nutritive components have been processed out and replaced with just a fraction of what was originally in the food, disease will subsequently develop.[2]

It is not just that Western societies are eating so much food – they are of course – but they are also eating foods from which "life" has been removed, because the ordinary valve nutrients of the food have been lost. Most foods we purchase from our grocery stores or supermarkets are indeed nutrient-deficient; they are not completely natural foods that will optimize human health.[3]

According to Mike Adams, the Health Ranger,

> In the food industry today, there is no official definition of "all natural". Today it can mean whatever the food manufacturing companies want it to mean. It can even mean, for instance, that all the chemicals found in the product are simply not listed on the food label! This is all perfectly allowed and tolerated by the American Food and Drug Administration (FDA) as well as by virtually every media outlet in the world. Cable news stations, magazines, newspapers and media giants are all too happy to take money from junk food manufacturers and run their advertisements claiming their foods are "all natural". There is absolutely no effort being made to determine whether such claims are true or even partially credible. Media companies simply take the money and run the promotions, regardless of whether such promotions tell the truth.[4]

Thinking Twice About Food and the Benefits Promised

Apart from choosing and buying food, it is important that we be more informed about the foods we consume. We should choose foods that *really* contain the most nutrients and are *really* the healthiest to eat. We need to eat natural plant foods that are packed with antioxidants and other essential nutrition-dense nutrients, including dietary fibre. Food comes in a "package deal". Therefore, the more compact, dense and nourishing a specific food group is, the healthier will be the cocktail of vital, life-saving and healthy nutrients that we consume.

Choosing Foods that *are* Nutrient-Dense

When nutritionists and dieticians speak about the "density" of foods, they are referring to **the amount of nutrients contained in the food divided by the number of calories**. In other words, foods are dense in nutrients when they have a lot of nutrients and do not contain many calories. It was Dr. Joel Fuhrman who invented "The Health Equation" which is:

H=N/C, or Health = Nutrients/Calories

The Health Equation was first published in 1999 in his book, *The Health Equation*, and described in more detail in a subsequent book, *Eat to Live*, wherein Dr. Fuhrman describes how the quality of calories impacts health. This equation means that your future health can be predicted by the micronutrient-per-calorie density of your diet.

Micronutrient-per-calorie density is important in devising and recommending menu plans and dietary programmes that will be the most effective for tackling weight loss and for preventing and reversing diseases.[7]

As Dr. Joel Furhman stated,

> Though micronutrient density is a significant dietary factor, it is not the only factor that determines health. For example, Vitamin D levels, B12, and proper omega-3 intakes are influential for optimal long-term health, as well as avoidance of sodium and other toxic excesses. These concerns are not addressed in the H = N/C equation. However, if the main focus is on consuming more micronutrient-rich natural foods, then the other important nutritional benefits automatically will follow: such as lesser sodium uptake, reduced calories, high fibre volume, lower saturated and cholesterol intake, a paltry glycemic index and an elevated satiety and phytochemical uptake.[8]

To recapitulate this basic idea, we can conclude that plant foods consistently contain a towering nutrient-dense score; which will be outlined and proven later in this book. This is since plants, furthermore contain vast amounts of antioxidants and abundant amounts of nutrients per calorie. On the contrary, animal, processed and refined foods, for the most part, consist of low levels of nutrients and are high in calories. This is why plant foods always win the prize for nutrient excellence and calorie reduction.

One must remember that fats and alcohol contain more calories per gram than carbohydrates and proteins, as the following table shows.

1 Gram	Calories
Fats	9
Alcohol	7
Carbohydrates	4
Proteins	4

There is no comparison between animal-based foods and plant-based foods. Plant foods always win the prize for food density and nutritional excellence. We cannot get the perfect arrangement of protective nutrients from industrialized refined/processed and animal foods – these foods surely lack many necessary health components and many of their vital nutrients have also been stripped away or destroyed during industrial preparation processes. The human body is starving for nutrients, and the best source of essential nutrients is retained, without any doubt, in plant vegetation.

Refined and processed foods contain additional added sugar, saturated fats, salt, artificial colors and preservatives that are unfit for human health. Tinned, packaged and preserved foods are very convenient and easy to prepare, but for health reasons, we should make sure that our daily food choices are obtained mainly from natural whole-rich foods. We should stay away from processed/refined foods as much as possible, and if we *do* choose to buy such foods, then we should make sure we thoroughly wash them in-order to help get rid of excess salts, pesticide residues and other additives or food preservatives.

Earth worthily, try to select and eat foods that come as natural as possible in nature, as it is these foods, that furthermore, contain the most nutrients and provide the most health benefits.

Notes for chapter 4

1. Hippocrates was born in 460BC and died in 360BC. The meaning of this dictum is the same as this ancient Ayurvedic proverb: "When the diet is wrong medicine is of no use. When the diet is correct medicine is of no use".
2. Fuhrman, J. (2012). *Raw for thirty video*. Retreived from http://www.rawfor30days.com/VideoSeries/dr-jorl-fuhrman/
3. Ibid.
4. Commentary by Mike Adams, the Health Ranger (2007). All Natural Foods. Retreived from http://www.counterthink.com/All_Natural_Foods.asp
5. Harris, W., *The Scientific Basis of Vegetarianism* (2005)
6. Fuhrman, J. (2011). *Eat to Live: The Amazing Nutrient-Rich Program for Fast and Sustained Weight Loss*. (revised ed). Little, Brown and Company.
7. Fuhrman, J. (2013). *How to Live, For Life: Prescription for Improving and Maintaining Great Health*. www.drfuhrman.com
8. Ibid.

5

When Good Men do Nothing!

All that is necessary for the triumph of evil is that good men do nothing.

Edmund Burke (1729-1797) born in Dublin:
*Author, orator, political theorist,
and philosopher*

I admire the man who is not scared and has the spirit to speak out against the "Controversial Inflammatory System of Control". This system of control is arresting and contesting many Naturopathic treatments from being applied in medical care treatments for diseases. The name of this man is Ty Bollinger. Indeed, every person concerned about improving health and reversing cancer should read his book, *Cancer: Step outside the Box*.[1]

Ty Bollinger writes and lectures on the cancer industry, health and lifestyle medicine. He strongly emphasizes that the cancer industry and the medical mafia are keeping vital and important information in relation to alternative life-saving cancer treatments from the general public in the U.S.A. He strongly advocates alternative cancer treatments which are far more superior to today's conventional treatments, and which could have in the past and at present save millions of lives worldwide.

The interpretation of Edmund Burke's quote, is that whenever good men or women do nothing, evil surely triumphs. Evil, wrongdoing, and sinful men must be opposed. God in the Bible actually commands those who are good to not just avoid evil, but to actively oppose it. So, when decent persons do nothing, they are no longer "good". One is not pleasant merely by doing nothing wrong, but he is good because he actively works for what is right and valid. "Let him eschew evil, and do well" (1 Peter 3:11). Further, the apostle James explains: "Therefore, to him that knows to be good, and doeth it not, to him, it is sin" (James 4:17).[2]

This quotation from Wayne Greeson is pertinent:

> "Be not deceived; God is not mocked; for whatsoever a man soweth, that shall he also reap" (Gal. 6:7). Those who fail or refuse to do good in the face of evil are sowing some dangerous seeds. They are doing nothing good as Jesus commanded them to do; they are helping evil to win and have ceased being good and have become partakers of the evil they did nothing to stop".

Of-course not all whom work in the medical care industry or follow a religion, are holding back on these alternative cures and sinful doings. That would be absurd to hold. Most religious believers, doctors, nurses and other health care professionals truly care for people. They are doing what they truly, honestly believe is paramount for their ill patients. However, what they have not learned in college, university or at medical school, they obviously cannot successfully teach or do to improve health in medical practice.

So Who *Will* Protect the Public?

Not governments, whose health systems have not been working and will never work, unless their system reboots and adjusts over towards a new approach to the treatment of disease.

Not the pharmaceutical industry, which pockets billions of money annually from chronic illnesses.

Not hospitals, whose livelihood certainly depends on ill patients.

Not the medical profession, in which physicians, doctors and nurses receive virtually no training in behavioral modifications and in nutrition and are handsomely rewarded for administering drugs and employing technical expertise.

Not the insurance industry, which profits by selling plans to the sick and, finally,

Not our many research funding institutions – too often they focus on biological details, such as individual nutrients that can be exploited commercially for profit.

I will fight for my own health and for your health. I will not be one of those men who stand back and do nothing. I started writing this book to verify and reveal the truth about these wrongdoings. I am truthfully acting and doing something. And so should we all. Most people are brainwashed by television ads, about so-called health food commercials, drug advertisements and cosmetic beauty products, which are detrimental towards health. We need to go against the grain, and not just sit around and carry on with our daily lives by condoning these facts, while billion-dollar corporations take control of our health.

The Public Needs New Avenues

The primary focus should be on whole foods, plant-based nutrition, and lifestyle changes. The public needs new avenues to understand this message, and that is why this book is so important. As already mentioned, a seismic revolution in health will not come from a pill, procedure or operation. It will occur only when the public is endowed with nutritional literacy, the kind of knowledge portrayed and highlighted in this book and in many other books about the advantages and health benefits that are found in eating a plant-based diet.

I truly would like all readers of this book to make the transition to a plant-based diet. This may loom like a challenge, however, it is only because so many of us have become addicted to diets that are high in salt, sugars, dairy products and fat, and the opiate-effects of other specific foods that the adventure of eating foods that contain few of these ingredients seems so difficult. Actually, it is not.

Most clinical scientists now clearly recognize, from scientific literature and practice, that such addictions can be resolved and reversed in a matter of weeks. This may only be fully accomplished when we start to consume healthier foods and start a detoxification program: relieving food additions and ridding ourselves of our first-born toxins. The first few weeks will be the hardest, but as soon as we get through this transition period, the rest will become easier. Once we see the health implications involved, we will certainly try to continue this dietary regime and lifestyle change.

Ty Bollinger is right! We need to "Step outside the Box". The body is an amazing machine. We *can* take control of our own health. Not be fooled by Tv ads or food commercials. All the human body needs is self-care and self-healing.

The best person to help the body accomplish these healthy goals is not your physician, nor any government health care system, but yourself.

However, only an experienced health care professional or health care system, whom strongly believes and truly acknowledges the healing powers of plants, that primarily focus on strategies that can educate and empower people to improve their diet and lifestyle, may be successful. Only those persons or systems whom worthily success in their brilliance in accomplishing these tasks, may someday decrease the rate of non-communicable diseases in their populations.

Notes for chapter 5

1   Bollinger, Ty. (2006). *Cancer Step outside the Box*. Infinity 510^2 Partners.
2   Greeson, Wayne. (Tuesday 12th, 2010). *When good men do nothing*. Adapted from Padfield.com article. Retreived from http://campbelllitigation.com/index.php?option=com_content&task=view&id=86&Itemid=47

6

What is a Plant-Based Diet?

There is a right diet for humans, and it is based on one key principle: eat plants.

Dr. John A. McDougall
American Physician and Author on Disease Prevention

The normal food of man is vegetable.

Charles Darwin (1809-1882)
English naturalist and geologist, best known for his contributions to evolutionary theory

A plant-based diet is a very simple one. It consists of avoiding anything that came from a source that ever had a face or a mother: "Eat what is grown, not born!"[1]

In other words, try to avoid all animal flesh (including fish), dairy, and eggs. What you do eat are the very finest foods that Mother Nature offers: green-leafy vegetables and other types of vegetables, fruits, nuts and seeds, including legumes. A healthy diet should consist of whole foods.

This means avoiding as much as possible refined foods: such as white rice, pasta and bread, oils, and staying away from artificial foods with chemical additives. In other words, try to center your food choices from whole, unrefined plants. Plants are natural, colorful and denser enrichment foods that contain abundantly more nutrients: foods that overall help, replenish and boost "peoples" health.

A well-structured, plant-based diet will meet all your nutritional needs – for calories, complex carbohydrates, essential fats, proteins, vitamins, minerals and other fundamental nutrients – without calorie counting, measuring nor worrying about portion sizes – just eat healthy plant foods whenever you are hungry. You will also lose weight naturally, without taking any form of pills or enrolling in any form of dietary plan. It is the easiest way to lose weight or to avoid gaining extra calories, and at the same time you *will* be eating healthier and improving your health.

A Food Guide Pyramid for a Plant-Based Diet should consist of:

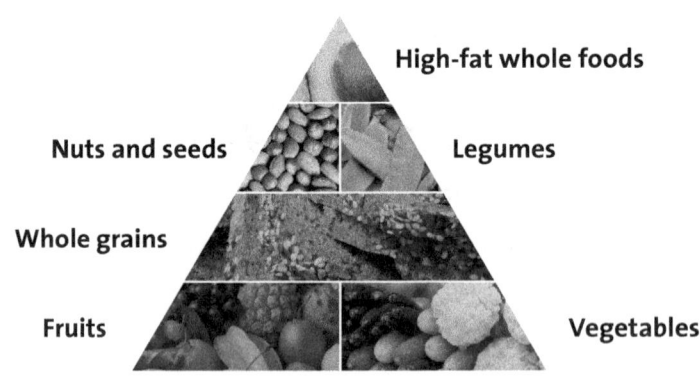

A healthy-food pyramid should focus upon a well-structured plant-based diet. A daily diet based mostly of vegetables and fruits, whole grains, legumes, nuts and seeds is best. There should be no, or very rare, intakes of animal-based foods, including refined oils and sugars, including dairy products. Processed and refined foods should be slowly eliminated from your diet.

Dr. Douglas N. Graham recommends that a healthy diet should consist of plant-based foods comprising of 80 percent complex carbohydrates, 10 percent protein, and the final 10 percent consisting of essential fats: "An 80/10/10 diet".[2]

The Five Essential Plant Food Groups:

1. Vegetables
2. Fruits
3. Whole Grains
4. Legumes (lentils, peas and beans)
5. Nuts and Seeds

To the above list one should always try to add two daily sources of supplementations, which should include:

1. The first being an adequate source of **Vitamin B$_{12}$** (unless you choose B12 fortified products), and
2. The second being **Vitamin D$_3$** (if you rarely get any exposure to direct sunlight).

As you may have noticed, the recommended foods *do not* include meat, dairy products, eggs, or refined/greasy fried goods. Why? It is because these are detrimental food choices and can cause the development of chronic diseases in humans, as you will read and has been fully revealed by scientific evidence.

The above nutritional dietary recommendations apply to all, especially to vegetarians, vegans and "too" pregnant or nursing women.

Diet and Lifestyle: Getting the "Right" Message Out!

Many people with cancer report feeling that if food plays a role in cancer, then they are somehow to blame for their disease. Blame and guilt often become concerns for people dealing with cancer. However, such guilt feelings are not helpful, and it makes no sense to blame oneself if one had no way of knowing that diet has a fundamental role in cancer prevention. Until major public education programmes, information and awareness are provided, including programmes spreading the word about the role of dietary factors that help people to change their lifestyle and bad eating habits, then cancer *will* still prevail and remain an epidemic in westernized societies.[3]

We need not only boost our own strength against cancer, but also break the chain or cycle of events for the next family generations. As already mentioned before, 1 in 8 women will contact breast cancer in their lifetime. There can be no fooling around with Cancer. Women need to take their diet seriously, especially if they already have Breast cancer. If persons cannot obtain these vital facts about this disease, then indisputably they will not be able to decide for themselves what to do next.

Why would the leading killer of young women not be the subject of prevention campaigns? Unfortunately, public attention is focused on diagnosing breast cancer by mammography and self-examination, not on preventing it through diet and lifestyle changes.[4]

In past years, and in most countries, there has been designated a "National Breast Cancer Awareness Month". Many women with breast cancer support this campaign, and they do well to march in the streets to raise funds and promote early prevention.

Television programmes and magazines, pick up the story. Nevertheless, each year, the press focuses mainly on mammography and self-examination. While such educational campaigns have merit, unfortunately information and awareness on the prevention of breast cancer through diet is completely left out. What the press in some countries does not know is that "National Breast Cancer Awareness Month" is sponsored by Imperial Chemical Industries (ICI), the makers of Tamoxifen: the anti-oestrogen drug used in the treatment of breast cancer.

The incidence of breast cancer cannot be reduced by early detection because mammograms and self-examination can only find existing cancers and are intended to make treatment more successful. Diagnosing cancer by mammograms or other cancer diagnostic tools will never be as effective as avoiding cancer entirely.[5]

Meanwhile, a tragedy is unfolding in Japan and in other Asian and westernized nations. The U.S. government has been pushing Japan to accept American agricultural products, particularly tobacco and beef. As demand for these deadly products declines domestically, sales are propping-up overseas. There is no shortage of patrons for Kentucky Fried Chicken, Burger King and McDonald's in Tokyo, Osaka, and other parts of the world. Meat, poultry, and egg consumption have increased eightfold; dairy consumption is fifteen times what it was in 1950. Fat intake in Japan climbed from less than 10 percent of calories in 1955 to 25 percent in 1987. Today these foods have expanded in western society's dietary regimes, and we have seen a dramatic rise of non-communicable diseases within their populations.[6]

Meanwhile, consumption of rice and green and yellow vegetables has dropped dramatically. Breast cancer death rates are increasing steadily: along with cancer of the colon, the ovaries, the prostate and the pancreas, among others.

As the consumption of more meat, dairy products, and refined foods assaults women's bodies and the origination of vitamin-rich vegetables and fruits are neglected, the effects are all too apparent: *an altered hormonal function, an unnatural age of puberty, and an impairment of the immune system* that could, otherwise, knock out cancer cells.

The Protective Power of Anti-Cancerous Fruits & Vegetables

Thousands of human and animal epidemiological studies have been reviewed and documented on the relationship between fruits and vegetables, and cancer.[7]

Many of these studies have conclusively proven that fruits and vegetables are potentially beneficial in fighting the battle against cancer in reducing cancer proliferation and growth. There are numerous food compound substances in fruits and vegetables that are protective, so the entire effect is not likely to be due to a single nutrient, antioxidant or phytochemical. A synergistic effect is required.[8] Clinical studies by Steinmetz and Potter have shown the human protective effects of a greater intake of fruit and vegetables, are effectively consistent too many types of cancer. The alliums' family of vegetables (garlic, onions, leeks, and scallions)[9] are particularly potent against cancer, while the cruciferous vegetables (cauliflower, cabbage, kale, broccoli, Brussels sprouts, etc.), contain sulforaphane and indole-3-carbinol which also have anti-cancerous effects.[10]

Do conventional dietitians, nutritionists or doctors know about this? It is a shame, but, unfortunately most of them do not!

There have been many other studies that have not shown the same beneficial cancer effects with plant foods. This could be due to the fact that not enough types or not the right amounts of anti-cancerous fruits and vegetables were directly tested in clinical trials. Food selection or quality bias against these studies might altogether account towards these erroneous results.

Many studies on the benefits of fruits and vegetables vis-a-vis cancer were usually in the range of what the American omnivorous diets normally eat: intakes of 4.5 and 6.2 servings per day or intakes of 4.3 and 5.4 servings per day[11]. But, there are other studies which showed and proved that a higher intake of fruits and vegetables (6.0 fruits, 11 vegetables, in addition to green powder from the juice of barley leaves, etc.)[12] that is equivalent to approximately another 100g per day of fresh dark greens, substantially reduced cancer growth. This means that it is highly probable that the range of intakes of fruits and vegetables performed in case-control studies and prospective population-based studies **do not** have a wide enough intake at the upper end: to detect the true possible impact of a very high intake of fruits and vegetables on cancer reduction.

A joint report by the American Institute of Cancer Research Fund and the World Cancer Research Fund found that a high fruit and vegetable diet reduces cancer of the mouth, the pharynx, the oesophagus, the stomach, the lungs, the colon, the rectum, the larynx, the bladder, the prostate and the breast.[13]

We need to open up our horizons to the fact that the more plant-based foods we consume: fruits and vegetables, nuts, seeds and legumes, the healthier we will be. It is not just about the quantity, but also, the quality of these specific plant foods that will work synergistically as anti-cancerous.[14] So, these food choices and changes in the way we look at "Diet vs. Disease" are vitally essential for the prevention of chronic-related diseases, including cancer.[15]

The American Institute of Cancer Research 11th Annual Research Conference on Diet, Nutrition and Cancer has concluded that: "By integrating or combining chemo-preventive approaches, from the use of single nutrients to multiple dietary constituents and functional foods, the scope of future cancer prevention strategies will be broadened".[16]

Research on eating behavior and changes in food patterns must be included in any cancer prevention strategy. A new paradigm for diet, nutrition and cancer prevention can be developed. How? By using multidisciplinary approaches that include lifestyle and environmental changes, dietary modifications and physical activity awareness to reduce the burden of cancer not only for high-risk individuals, but for the general population as well. A healthy diet can diminish our risk and, at the same time, build the human body's defenses. To the extent that such a diet and other lifestyle changes are put to work, we can hope to turn the tide of this epidemic disease (breast cancer) among women and respectively for other chronic-related diseases in today's "Late modern period."[17]

Notes for chapter 6

1   Caldwell B. Esselstyn, Jr., M.D. Foreword by T. Colin Campbell, Ph.D., author of *The China Study*. (2007). *Prevent and Reverse Heart Disease*. Avery Publications.

2   Douglas N. Graham. (2006). *The 80/10/10 Diet* (1st ed). Published by FoodnSport Press.

3   Barnard, N. (1993). *Food for life: Cancer and Immunity*, **3**, pp.67-68.

4   Ibid.

5   Barnard "Food for life", 1993.

6   Waynder, E.L., *et al.* (1991). Comparative epidemiology of cancer between the United States and Japan. *Cancer*, **67**: pp.746-763. Barnard "Food for Life", 1993.

7   Kagawa, Y., *et al.* (1978). Impact of Westernization on the nutrition of Japanese; changes in physique, cancer, longevity and centenarians. *Prevention Medicine*, **7**, pp.205-217.

8   Heber, D. *et al.* (2004). Phytochemicals beyond Antioxdation. *American Society for nutritional Sciences. Journal of Nutrition*, **143**; 3175S-3176S.

9   Steinmetz, K.A., Potter, J.D. (1996). Vegetables, fruit, and cancer prevention: a review. *Journal Am Diet Assoc*, **96**, pp.1027-1039.

10 Ibid.

11 Fleischauer, A.T., Poole, & C., Arab, L. (2000). Garlic consumption and cancer prevention: meta-analyses of colorectal and stomach cancers. *American Journal of Clinical Nutrition*, **72**, pp.1047-1052. Zhang, S.M., *et al.* (2000). Intakes of fruits, vegetables, and related nutrients and the risk of non-Hodgkin's lymphoma among women. *Cancer Epidemiol Biomarkers Prev*, **9**, pp. 477-485. Michaud, D.S., *et al.* (1999). Fruit and vegetable intake and incidence of bladder cancer in a male prospective cohort. *J National Cancer Inst*, **91**, pp.605-613; Cohen, J.H., Kristal, A.R. & Stanford, J.L. (2000). Fruit and vegetable intakes and prostate cancer risk. *Journal National Cancer Inst*, **92**, pp.61-68. Kolonel, L.N., *et al.* (2000). Vegetables, fruits, legumes and prostate cancer: a multiethnic case-control study. *Cancer Epidemiologic Biomarkers Preview 2000*, **9**, pp.795-804. London, S.J., *et al.* (2000). Isothiocyanates, glutathione S-transferase M1 and T1 polymorphisms, and lung-cancer risk: a prospective study of men in Shanghai, China. *Lancet*, **356**, pp.724-729.

12 Willett, W.C., *et al.* (1999). Fruit and vegetable intake in relation to risk of ischemic stroke, *JAMA*, **282**, pp.1233-1239.

13 Donaldson, M.S. (2001). Food and nutrient intake of Hallelujah vegetarians. *Nutrition & Food Science*, **31**, pp.293-303. WCRF/AICR. (1997). Food, Nutrition and the Prevention of Cancer: A Global Perspective. *World Cancer Research Fund / American Institute for*

Cancer Research. Fleischauer, A.T. *et al*. (2001). Garlic and cancer: a critical review of the epidemiologic literature. *J Nutrition,* **131**, pp.1032S-1040S.

14 Heber, D. *et al*. (2004). Phytochemicals beyond Antioxidation: American Society for nutritional Sciences. *Journal of Nutrition,* **143**, pp.3175S-3176S.

15 Donaldson, M.S. (2004). Nutrition and Cancer: A review of the evidence for an anti-cancer diet. *Journal of Nutrition,* **3**(19), pp.1-21.

16 Liang, V. *et al*. (2001). Diet, Nutrition and Cancer Prevention: Where are we going from here? UCLA Center for Human Nutrition, Los Angeles, CA and *American Institute for Cancer Research, Washington, DC. *J. Nutr,* **131**, pp.3121S–3126S.

17 Ibid.

7

The Disappointment of Conventional Treatments

*Health is not valued
until sickness comes.*

Thomas Fuller (1608-1661)
English churchman, historian and prolific author

Most people will agree that eating a diet high in plant foods will prevent most cancers. The same people, however, will find it difficult to believe that they can actually reverse cancer by adopting a plant-based diet.[1]

No one with a splitting headache or back pain is actually suffering from a deficiency of aspirin. And no one with atherosclerosis is suffering from a deficiency of statins (drugs that help reduce cholesterol levels). Similarly, no one with cancer has a deficiency of chemotherapy and radiation. But while these therapies might temporarily relieve the symptoms of the burden of the disease, they do not focus on, attack or treat the underlying cause of the disease.

For example, a woman might be afflicted with metastatic breast cancer because a few years back, she lost a child and this still drives her catecholamines into stress mode and releases excess cortisone – as often happens when a person is under stress. Indeed, this could lead to hypertension, and a depressed immune system that affects other body ailments. She would probably go to bed eating a box of high sugary candy bars each night. She is probably also deficient in essential fatty acids, zinc, Vitamin D and E, and she has an imbalance of oestrogen and progesterone in her body – initially due to consuming too many animal products and not eating adequate amounts of nutritious plant foods. Her diet is lacking necessary nutrients, and she is not looking after herself properly, basically owing to the tragic loss of a child. The main cause of her disease is the loss of the child and her depressed mood – that has skyrocketed her stress levels – that may have led to her wrong lifestyle and eating patterns.

Her oncologist could remove her breasts, give her Tamoxifen to reduce oestrogen levels, and apply chemo and radiation. However, none of these therapies deals with the *underlying* cause of her condition. Her breast cancer may obviously return, unless the real driving forces of the disease are reversed.

The way forward and best medical treatment for such a mental and physical disturbed woman would most observably be too:

1. Give her professional guidance to help her cope with the loss of her child and offer her hope of a happy, sustainable life.
2. Start her on a lifestyle medicine management plan to rebuild her immune system and to improve her overall health. This will help her stress levels to return to normal; her eating habits will improve and which, indeed, may counteract the conditions that had caused her cancer to grow and metastasize in the first place.
3. Find friends and family that will help and support her, by respecting her wishes in starting a good and healthy balanced diet with recommended physical exercises incorporated with stress-reducing techniques to help her reverse her cancer.

Removing her breasts will only make her condition worse. Mastectomy would surely make her more depressed and may destroy all her hope of retaining a happy and normal life. Conventional treatments may sound negative, but these are the facts of the matter. This may sound negative, but this is the fact of the matter.

Where is conventional medicine heading? Obviously, it is heading in the wrong direction. One does not have to be a genius to realize that this woman desperately needs diet & lifestyle guidance with psychological support, *before* her oncologist should even think of treating her condition with unnatural and toxic disease-causing treatments: Chemotherapy, radiation and surgical procedures.

Scientific evidence has long proven that a healthy diet incorporated with daily amounts of exercise can reduce our risk of acute and chronic-related diseases, including cancer.[2]

In this sense, it is pertinent to remember the case of Nathan Pritikin, who was the first person who proved that heart disease *can* be reversed with a plant-based diet.[3] At that time, mainstream medicine had held that this was simply impossible! After Nathan Pritikin reversed his own heart disease, he was greeted with 20 years of scepticism and doubt by the medical profession. To prove his point, he willed that his body be autopsied, and the results published in a medical journal (The New England Journal of Medicine, 4 July, 1985). The autopsy proved that his arteries were as clean as a teenager's.[4]

Through the work of Dr. Dean Ornish and Caldwell B. Essellstyn, Jr, and others, it has now been proven a thousand times over that Nathan Pritikin's heart disease reversal was not a spontaneous remission, but the result of the adoption of a plant-based diet. However, despite such overwhelming evidence, hospitals still have not diverted to nutrition as the primary method of treating heart disease. Why? There is simply no money to be made from it.[5]

Alas, conventional medicine continues to treat the symptoms of heart disease with ineffective stents, statin drugs and bypass surgeries – all of which bypass the main cause of the ailment.

Another case of treating symptoms is Adult-Onset Diabetes (AOD). The American Diabetes Association will tell you that there is no cure for AOD. What they really intend by this is simply that "there is no drug that can indeed cure the disease". But here too, AOD can be easily and quickly cured through a change in diet and lifestyle.

Dr. John. A McDougall (Dr. McDougall's Health and Medical Centre) wrote:

> I don't really believe there aren't intelligent people in the scientific community, which have any doubt that Type II Diabetes is not caused by the Rich-Western Diet. I think everyone agrees with that. Nevertheless, what they do not do is actually not take the ultimate next step, and say, "OK, this is the cause of AOD " and so what we all should do is try to change the diet.[6]

Brenda Cobb (Living Foods Institute):

> Patients come into our clinic and get on our program, and within 24 hours they have cut their dose of Insulin by 50 percent. On day 5 or day 6, they do not need insulin injections at all, ever again. Type II Diabetes can be reversed within a week on a plant-based diet. When you hear these true medical stories you think, "That's impossible". The doctors say that it is incurable, and that it cannot be reversed. However, that is not true. You can reverse Type II Diabetes. You can *completely* heal or control diabetes, and very, very quickly. It is not only Diabetes, but also Lupus, Multiple Sclerosis, Parkinson's disease, Chronic Fatigue Syndrome, heart diseases, obesity, autoimmune diseases and cancers,

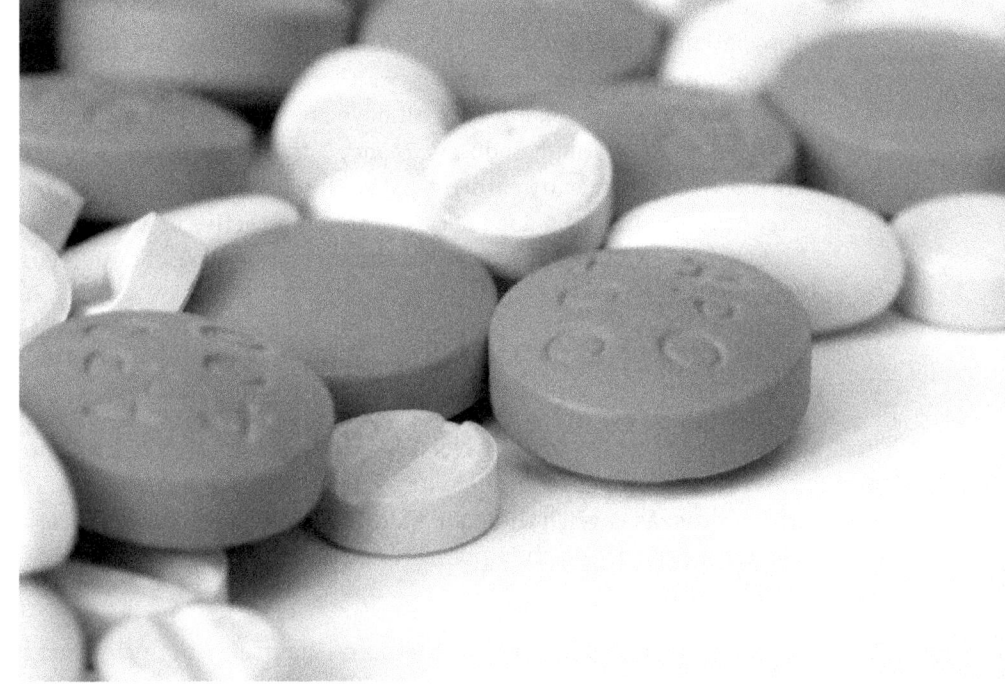

etc.... Patients, include people with Lupus, sometimes even people who come in wheelchairs and patients who cannot even raise their arms high enough to feed themselves. Within 10 days, patients are up and out of their wheelchairs and walking steadily on their own accord.[7]

This is corroborated by Charlotte Gerson, of The Gerson Institute: "In most cases it only takes 5 to 7 days. In a few, particularly severe cases of diabetes, it might take up to two weeks to reverse the disease."

Similarly, Dr. Thomas Lodi:

We find that 5 to 7 days are enough for someone with Type II Diabetes. However, if the patient's diet is based on consuming 100% uncooked, raw, plant-based foods, they no longer need insulin injections. I have not seen that not happen yet.[8]

Dr. John McDougall too agrees:

So, yes, the emphasis of the treatment by the American Diabetic Association, by the American Heart Association, and by the National Osteoporosis Foundation and other health and research facilities is in the erroneous direction. But they all have one thing in common: they all have a lot of "Hugh Industry Money", that keeps them focused on the wrong medical treatment path. This just did not happen by circumstance. This happened because there is an active flow of money that keeps them in the direction of drugs and not in the right direction of a good diet.[9]

Major cancers may be prevented with lifestyle medicine: a plant-based diet, exercise with and combined too other non-drug modalities.

Just as drug companies with investing interests have blocked the prevention of heart disease, diabetes, systemic lupus, Parkinson's and other chronic diseases with diet, the said health and research centers have also done everything they possibly could to discourage the use of diet to treat cancer.

During the early 1900's allopathic medicine has seen many natural cancer cures: involving the intervention of dietary plant foods as a threat to the bag of profitable toxic treatments. As I have already mentioned, our body is curing cancer every day of our lives, and the difference between prevention and reversal is one of degree, not of kind. In fact, the average American has six serious cancer episodes during their lifetime, that is, the human body can actually reverse serious cancers about 6 times during their life without them even realizing it. Those with healthy immune systems do not know the battle their bodies are waging. Unfortunately, those with weakened immune systems will find out about these battles in hospital. The foundation of a healthy immune system is diet. A good diet enables us all to reverse cancer every day of our lives.[10]

Despite the heroic-sounding "war on cancer" cry, and the "destroy the enemy" rhetoric, we must remember that the body is not a battlefield, but a complex and sensitive system of self-regulation with profound healing possibilities. The immune system is incredibly organized, and can easily stimulate our brain in terms of complexity, subtlety and self-awareness. Even so, conventional treatments do everything to destroy this masterpiece![11]

In other words, the underlining cause was not treated. Microscopic tumours can still be invisibly looming or hidden inside tissues or organs. Not all tumours can be detected by conventional procedures. Nevertheless, we send patients back home, without educational knowledge on how and why the cancer had emerged in the first place. Thus, cancer patients continue to poison themselves with the same foods and lifestyle that had developed their tumours in the first place. Cancer may then re-emerge with a vengeance.[12]

We are indeed selling cancer patients out, because they think they have been cured from their disease but, in reality – and sad to say – they would just be on another long journey to more family sorrow, stress, suffering and unnecessary pain.

Why are Most Doctors Against Alternative Medical Treatments?

Many doctors are against alternative medical treatments. Maybe because from their first day at medical school, doctors are convinced and educated that chronic diseases' may only be treated or reversed by conventional treatments?

Your general practitioner (G.P.) has probably been brainwashed and persuaded that the only cures or controllable treatments for cancer are chemotherapy, radiation and surgical medical procedures. Most doctors also believe that there is no cure for cancer, while in reality there are. Most doctors have a tendency to believe not only that what they were taught in medical school is valid and true, but also that what they have not been taught is unimportant! As a result of this, when it comes to cancer, most medical practitioners still think "inside the box". Sadly, those doctors who think "outside the box", and who treat the underlying cause of any disease rather than just the symptoms of disease, are sometimes labelled as "quacks".[13]

A medical health plan for doctors and for the medical establishment on how to treat cancer patients with diet and lifestyle changes is indispensable and urgently needed.

More importantly, doctors and the medical establishment need to focus more on educating the general public about lifestyle medicine, and not on treating cancer patients with chemotherapeutic drugs, surgery and radiation that do not treat the underlying cause of any chronic disease. A healthy diet, exercise and lifestyle should be, without any doubt, their primary goal in reversing chronic diseases like diabetes, heart diseases, autoimmune diseases and cancers.

Heart disease is still the worldwide number one killer in Westernized societies. If the medical industry still focuses on chemotherapy, radiation and other diagnostic and surgical procedures to eradicate cancer, while still completely ignoring and does nothing to change their patients bad eating habits and lifestyle, unfortunately, and without any doubt whatsoever, cancer will prevail in becoming societies secondary pandemic killer. However, in the decades to come, cancer may well become the number one leading cause of death in the Western world, winning over heart-related diseases.

The Time to Act is Now.

Notes for chapter 7

1. Anderson, M. (2009). *Healing Cancer from inside out: A Practical Guide to healing Cancer with the RAVE Diet & Lifestyle*. (1st ed). Published by www.RaveDiet.com
2. Roberts, C. & Barnard, R.J. (2005). "Effects of exercise and diet on chronic disease". *Journal of Applied Physiology,* **98**, pp.3-30.
3. Monte, T. & Pritikin, I. (October, 1987). *Pritikin: The Man Who Healed American's Heart*. Hardcover, 102 pages. Published by Rodale Press, Emmanus, Pennsylvania.
4. Hubbard, J.D, Inkeles, S., & Barnard, R.J. (July 4th, 1985). Nathan Pritikin's Heart. *The New England Journal of Medicine,* **313**(52) DOI: 10.1056/NEJM198507043130119
5. Anderson "Healing Cancer from inside out", 2009.
6. McDougall, J. Dr. McDougall's Health & Medical Center. Retreived from www.drmcdougall.com/health/education/
7. Cobb, B., Living Food Institute. Retreived from http://livingfoodsinstitute.com/index.php
8. Lodi, T. *An Oasis of Healing.*
9. McDougall "Health & Medical Center."
10. Anderson, M. (2009). *Healing Cancer from inside out.* video disc 1.
11. Ibid.
12. Anderson, M. (2009). *Healing Cancer from inside out* Part 1: The Failure of Conventional Treatments: Saying Goodbye to the Cancer Industry, page 51. Published by www.RaveDiet.com
13. Bollinger, T. *Why is my doctor against Alternative Cancer Treatments?* Retreived from www.cancertruth.net

8

The Importance of Choosing the Right Form of Vitamin B_{12} (Methylcobalamin)

While getting enough vitamin B12 is an important factor for maintaining general well being, getting the right kind, in the form of methylcobalamin is equally important. In fact, choosing correctly could very well mean the difference between good health and disease.

Knox, K. (September 17th, 2009)
Natural News Article:
The Right Kind of Vitamin B_{12} is Vital for Treating Deficency.

Vitamin B_{12} is an essential compound of the human body and as like other vitamins, it is vital for various biological functions. But firstly, let us take a closer look at the mechanisms in how vitamin B_{12} is transported, absorbed and stored inside the human body.

Transport of Vitamin B_{12}

Gastric intrinsic factor (GIF) is a glycoprotein produced by the parietal cells of the stomach. It is necessary for the absorption of B_{12}, later on in the small intestine.

Before entry into the stomach, B_{12} binds to haptocorrin. The resulting complex enters the duodenum, then in the pancreas. Pancreatic enzymes digest haptocorrin in the less acidic environment of the small intestine, aiding B_{12} finally to bind to intrinsic factor. This new complex travels to the ileum, where specific epithelium cells endocytose them. Later, B_{12} dissociates once again to bind to another protein, transcobalamin II, and this new compound structure can exit the epithelium cells of the ileum to enter the liver. B_{12} is then circulated and used for normal metabolic chemical reactions.

Damage to the Parietal Cells of the Stomach Can Cause Health Complications

Pernicious anemia is an autoimmune disorder whereby auto antibodies directly attack the parietal cells themselves or intrinsic factor. This autoimmune disease may lead to an intrinsic factor deficiency, megaloblastic anemia, or subsequent malabsorption of vitamin B_{12}.

Atrophic gastritis can also cause intrinsic factor deficiency and anemia through damage to the parietal cells of the stomach wall. Pancreatic exocrine insufficiency can hinder with normal dissociation of vitamin B_{12} from its binding proteins in the small intestine, preventing its absorption via the intrinsic factor complex.

Other risk factors may also contribute to pernicious anemia: anything that damages or any surgical procedure that removes a portion of the stomach's parietal cells, including bariatric surgery, excessive consumption of alcohol, gastric tumours and gastric ulcers.

B_{12} Storage: Liver and Muscles

The average non-vegetarian stores between 2,000-3,000mgs of Vitamin B_{12} and loses about 3µgs per day. B_{12} is stored mainly in the human liver (60%) and the rest mainly in the muscles (30-40%). Through the bile, we secrete 1.4µgs per day in the small intestine and a healthy person should reabsorb about 0.7µgs.[1]

Due to the extremely efficient enterohepatic circulation of B_{12}, the liver can store several years' worth of vitamin B_{12}; therefore, nutritional deficiency of this vitamin is rare.

The two main functions of Vitamin B_{12}

1. Methylcobalamin is used by the enzyme methionine synthase to change homocysteine into methionine. When this enzyme is not functioning adequately it can substantially step-up homocysteine levels in the body, which studies have linked to an increased risk of heart disease, strokes, blood clot formation and to the deterioration of nerves and arteries.
2. B_{12} is enzymatic coenzyme is using 5-deoxyadenosylcobalamin in the enzyme methylmalonyl-CoA mutase in the conversion of methyl malonyl-CoA to succinyl-CoA, an intermediate of the citric acid cycle. Succinyl-CoA aids towards porphyrin synthesis and so towards the formation of human blood cells. This is one reason why people who are folic acid-deficient or B_{12}-deficiency can also develop haemolytic anemia, clinically called *megalobastic anemia*.[2]

The Homocysteine Dietary Connection

Blood levels of homocysteine tend to be highest in people who's diet consist of high amounts animal-protein and consume fewer fruits and green-leafy vegetables. Plant foods provide vast amounts of folic acid and other essential B vitamins that are necessary to help the body rid itself of excess homocysteine. Stress and coffee consumption are other factors that can also increase homocysteine levels. Coffee and stress-induced neurotransmitters – epinephrine and norepinephrine are metabolized in the liver via a process that uses methyl groups. This can also increase the need for folic acid. Elevated homocysteine levels may also be due to low levels of thyroid hormone, kidney disease, psoriasis and some medications.[3]

In addition to the dietary connection, homocysteine is also produced in the body from another amino acid, methionine. One of methionine's main functions is to provide methyl groups for cellular reactions. A methyl group is a chemical fragment consisting of one carbon and three hydrogen atoms.

When methionine donates a methyl group for a cellular reaction, it becomes homocysteine. Typically, homocysteine may then reserve another methyl group from either folic acid or vitamin B_6 to regenerate methionine. So, essentially, beside taking the best form of Vitamin B_{12}, which is methylcobalamin, we also need a good supply of B Vitamins, especially pyridoxine (B_6) and folic acid (B_9), in order to facilitate the mop-up of the active methyl groups left behind after the metabolism of methionine absportion, as this can reduce the risk of the development of increased levels of homocysteine. Methylation may also cause DNA damage. A multiple of the essential B Vitamins can be provided by consuming leafy-green vegetables, fruits, whole grains and legumes. That is one of many health reasons why eating plant-based foods are vitally imperative for the reduction of homocysteine levels.

FACT BOX

It is estimated that around 50 percent of vegans may actually be B_{12}-deficient.

Why is Vitamin B$_{12}$ Lacking in Plant Foods?

Many people ask me, "Well, if a vegetarian or vegan diet is a healthy one, and if it is in keeping with our natural relationship with nature, then why is vitamin B$_{12}$ lacking in plant foods, and only retained mainly in animal products?" My reply to this question is hereunder.

Animal products *do* contain Vitamin B$_{12}$ and, indeed, plant foods *do not*. Why? Animals ingest plants and drink water that feed the microorganisms that produce this vitamin. Vitamin B$_{12}$ is constantly being produced throughout the environment by **bacteria**.

If you were living in the wild, as our ancestors did, you would most likely get plenty of B$_{12}$ from the water you drink – although nowadays it is usually unhealthy to drink directly from rivers, streams or ponds, as these water sources are often contaminated. Instead, most of us drink water which has been distilled or chlorinated. However, if you were living the way our ancestors did, you probably would be ingesting B$_{12}$ along with the bits of dirt containing microorganisms found in the little grooves or on the peels of your fruits and vegetables.

Unfortunately, due to industrial production, nowadays our soils are sprayed with chemical fertilizers and pesticides, with the result that they are devoid of B$_{12}$, which was once abundant in plant foods. Furthermore, food today is so sanitized that even if there remains a small quantity of B$_{12}$ in the dirt in which veggies grew we still may not find it. The recommended dietary allowance for vitamin B$_{12}$ is incredibly tiny: just 2 micrograms per day, i.e. 2 millionths of a gram.[7]

Animal products do contain the best food sources of B$_{12}$, however, these are not the healthiest source because of the heavy baggage that comes along with them. It is like we cannot get the iron in beef without the saturated fat; the protein in pork without lard; the calcium in dairy without the casein and hormones. We humans, unfortunately, cannot obtain the B$_{12}$ from animals without also consuming stuff we do not want, like cholesterol.[8]

Vitamin B_{12} deficiency is common among persons eating vegetarian and particularly vegan diets because of failure to take B_{12} supplements or not consuming B_{12}-fortified foods.[9]

Kerri Knox, a Registered Nurse, writes as follows:

> Once someone has overcome all hurdles to get appropriate testing and is told that they have vitamin B_{12} deficiency, there is still a problem to overcome: getting the right **kind** of vitamin B_{12}. There are several kinds of vitamin B_{12}, but only **one**, *methylcobalamin*, is the correct form that should be used in the vast majority of cases. Yet few doctors use methylcobalamin and prefer to use their prescription pads and tradition over good science in their decision to supplement Vitamin B_{12}.[10]

Vitamin B_{12} is actually a generic term for a class of compounds called *the Cobalamins*. So, whenever a person has vitamin B_{12} deficiency, it is more scientifically called "Cobalamin deficiency". This makes perfect sense when you understand that the various formulations of Vitamin B_{12} all end with the suffix Cobalamin. While there are many forms of B_{12} or Cobalamins, only three are generally used as dietary supplements, namely:[11]

- Hydroxocobalamin
- Cyanocobalamin, and
- Methylcobalamin (which is the best active form of Vitamin B_{12} supplement)

Cyanocobalamin is probably the most commonly used B_{12} supplementation in the medical world. It is usually given as "B_{12} shots". But cyanocobalamin is actually the *worst* choice, even though doctors in the US are more likely to prescribe it over any other form. Not only does cyanocobalamin require a higher dosage for the same effectiveness of hydroxycobalamin, but it is "entirely ineffective" for several different conditions related to vitamin B_{12} deficiency.[12]

As such, it has been suggested repeatedly by several researchers, starting with Dr. A.G. Freeman in 1970, that cyanocobalamin should be removed from the market. While Great Britain followed through

THE IMPORTANCE OF CHOOSING THE RIGHT FORM OF VITAMIN B$_{12}$ (METHYLCOBALAMIN)

with research recommendations and removed this inferior product, doctors in the United States have no such restrictions and still use cyanocobalamin routinely.

While hydroxocobalamin is preferred over cyanocobalamin, another formulation called *methylcobalamin* is actually the best choice. Technically a "coenzyme" of vitamin B$_{12}$, it is almost never used despite being effective, readily attainable, inexpensive and available in both sublingual preparations and injectable fomulations. This is indeed a pity, because there are many people that could very well benefit from the methylcobalamin form of vitamin B$_{12}$ that would *not* otherwise benefit from the other forms.[13] Degenerative neurological problems are where methylcobalamin shows its greatest benefits over other cobalamin preparations, and it is often one of the *only* promising treatments of these tragic diseases.

While Japan uses methylcobalamin nearly exclusively, and it is the form present in prescription, vitamin B_{12} there, the United States have virtually ignored the hundreds of studies that show the benefits this simple vitamin can bring.

Methylcobalamin is physiologically equivalent to vitamin B_{12}, and can be used to prevent or treat pathologies arising from a lack of vitamin B_{12} (vitamin B_{12} deficiency), such as pernicious anemia.

Not only has methylcobalamin been shown to work in neurological diseases, but it also helps with the elimination of toxic substances in the body. One of the ways that humans detoxify is through a process called "methylation". Methylation is a *critical* function of a healthy body, but all too often we "use up" the necessary raw materials because of our nearly constant exposure to environmental pollutants.

Methylcobalamin is actually able to replenish the "methyl" portion that is missing in methylation, while the other forms of vitamin B_{12} *requires* a methyl donor in order to be converted into a biologically active form in the blood.[14]

Therefore, people who already have methylation detoxification problems, such as children with autism, can actually be made *worse* if other types of vitamin B_{12} are administered!

While getting *enough* vitamin B_{12} is significant for maintaining a general well being, getting the right *kind*, in the form of methylcobalamin, is equally important. In fact, a correct choice could very well spell the difference between good health and disease.

Medical laboratory evidence for the effects of vitamin B_{12} was seen in a study with Vitamin B_{12} inadequate rats. It was found that the colonic DNA of the B_{12}-deficient rats had a 35% decrease in genomic methylation and a 105% increase in uracil incorporation – or both changes that could increase the risk of carcinogenesis. Methylcobalamin has also been shown to increase survival time and reduce tumour growth in laboratory mice.[16]

Two prospective studies (Washington Country, Maryland and the Nurse's Health Study) have shown that in the case of inadequate (lower) B_{12} amounts in the human body (and not lack of or deficiency of B_{12}) there was a significantly higher risk of breast cancer. So, there is evidence from laboratory studies that show that B_{12} is an important vitamin for DNA repair, genetic stability, including carcinogenesis and cancer therapy.[17]

For vegetarians and vegans, it is important to include foods that are fortified with Vitamin B_{12} or supplementary sources of the vitamin, e.g. nutritional yeast powder, which is fortified with additional B-vitamins, including B_{12}. For vegan, pregnant or nursing women, B_{12} supplements are indispensable.

Methylcobalamin represents one of the best values in nutritional products.

So, always choose methylcobalamin supplements over the other two forms of B_{12}. By choosing the right supplement form of B_{12}, we all can benefit in maintaining a good source of this essential vitamin. Methylcobalamin is comparably cheap and it also has a wider array of potential-health benefits.[18]

Notes for chapter 8

1. Cousens, G. M.D. (June 7th, 2012). *The Importance of Vitamin B$_{12}$*. The Healers Journal. Retrevied from http://www.thehealersjournal.com/2012/06/07/the-importance-of-vitamin-b12/

2. Ibid.

3. Weil, A. M.D. *Dr. Weil's Condition Care Guide. Elevated Homocysteine.* Retreived from http://www.drweil.com/drw/u/ART03423/Elevated-Homocysteine.html.

4. Loehrer, F.M., *et al.* (1996) "Low whole-blood S adenosylmethionine and correlation with 5-methyltetrahydrofolate and homocysteine in coronary artery disease." Arterioscler. Thrombo-sis Biol., **16**, pp. 727-733.

5. Jacob, A.R. (2000). Folate, DNA methylation, and gene expression: factors of nature and nurture. *Am J Clin Nutr,* **72**(4), pp.903-904.

6. Gilsing, A.M.J. *et al.* (2010). Serum concentrations of Vitamin B$_{12}$ and folate in British male omnivores, vegetarians and vegans: Results from cross-sectional analysis of the EPIC-Oxford cohort study. *European Journal of Clinical Nutrition,* **64**. pp.933-939.

7. Robbins, J. (2001). The Food Evolution. Part 1: Food and Healing: A Healing Plant-Based Diet. (p.90-91). Published by Conari Press.

8. Greger, M. (February 6th, 2012). *Safest Source of B$_{12}$*, Volume 7 video. Retreived from http://nutritionfacts.org/video/safest-source-of-b12/

9. Greger, M. (August 27th, 2011). *Vegan Epidemic,* Volume 5 video. Retreived from http://nutritionfacts.org/video/vegan-epidemic/

10. Knox, K. (September 17th, 2009). *The Right Kind of Vitamin B$_{12}$ is Vital for Treating Deficiency.* Vitamin B$_{12}$. Natural News. Retreived from http://www.naturalnews.com/027045_vitamin_B12_cyanocobalamin_methylcobalamin.html

11. Guest, J.R. *et al.* (April 1964). Methycobalamin as a source of the methyl group of methionine: Vitamin B$_{12}$ Coenzymes. *Annals of the New York Academy of Sciences,* **112**, pp.774-790.

12. Faigin, R. (2005). *B$_{12}$ (Cobalamin).* Hormonal Fitness News #9, pp.1-5. Retrevied from http://www.hormonalfitness.com/facts/HFN%209%20-%20b12%20Cobalamin.pdf . Guest, "Methycobalamin", 1964.

13. Ibid.

14. Knox, *"The Right Kind of Vitamin B$_{12}$ is Vital for Treating Deficiency",* 2009.

15. Greger, *"Vegan Epidemic",* 2011.

16 Choi, S.W. *et al.* (2004). Vitamin B-12 deficiency induces anomalies of base substitution and methylation in the DNA of rat colonic epithelium. *Journal of Nutrition*, **134**, pp.750-755; Shimizu, N., *et al.* (1987). Experimental study of antitumor effect of methyl-B_{12}. *Oncology*, **44**, pp.169-173; Wu, K. *et al.* (1999). A prospective study on folate, B_{12}, and pyridoxal 5'- phosphate (B_6) and breast cancer. *Cancer Epidemiol Biomarkers Prev*, **8**, pp.209-217.

17 Shimizu "Experimental study", 1987. Zhang, S.M. *et al.* (2003). Plasma folate, vitamin B_6, vitamin B_{12}, homocysteine, and risk of breast cancer. *Journal of the National Cancer Institute*, **95**, pp.373-380. Donaldson, M. (October 20th, 2004). Nutrition and Cancer: A review of the evidence for an anti-cancer diet. *Journal of Nutrition*, **3**(19): pp.1-21.

18 Freeman, A.G. (November 1992). Cyanocobalamin – a case for withdrawal: discussion paper. *The Journal of the Royal Society of Medicine*, **85**(11), pp.686-687.

9

The Health Benefits of Vitamin D₃

Vitamin D is cholecalciferol, a hormone. Deficiencies of hormones can have catastrophic consequences.

Dr. William Davis
A Milwaukee-based American cardiologist and author of health books

Vitamin D is a fat-soluble vitamin that is naturally present or added in very few foods: fortified milk, cereal foods; oily fish such as salmon, mackerel, sardines and cod-liver oils. There are also small amounts of Vitamin D in seaweed and irradiated mushrooms.[1] A dietary supplement or food source of Vitamin D_3 is essential for human body functions and for breast cancer prevention.[2] It is produced endogenously. This happens whenever ultraviolet rays from the Sun (sunlight) strikes the skin and triggers vitamin D synthesis. Vitamin D obtained from sun exposure, food and supplement is biologically inert and must undergo two hydroxyls (hydroxyl group: OH) in the body for activation:[3]

1. The first occurs on exposure to sunlight, and converts 7-dehydrocholesterol on the skin to Vitamin D_3. The liver then

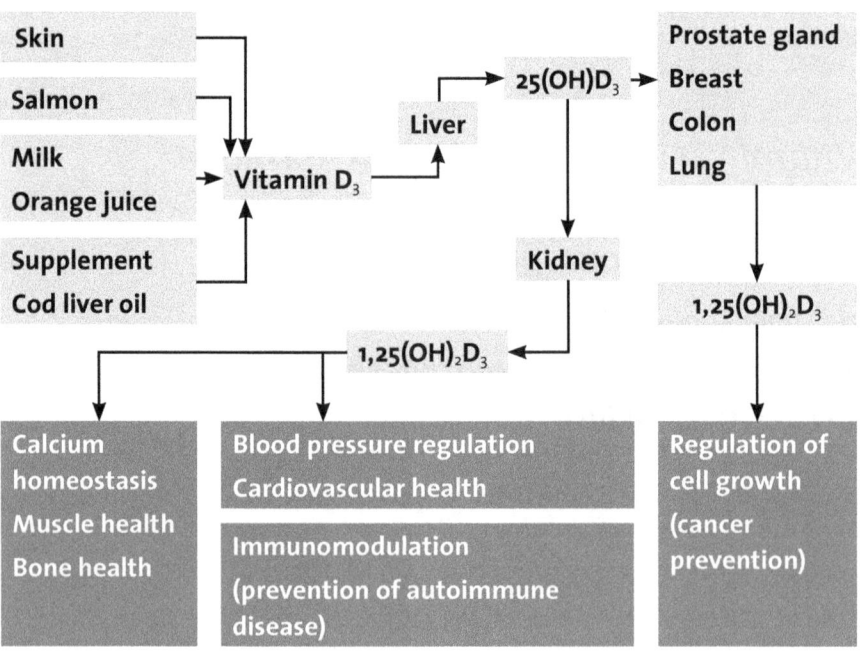

Metabolism of 25 (OH) D_3 to 1,25 (OH) $_2D_3$ in the kidney and other organs and the Biological consequences[1]

converts vitamin D to 25-hydroxyvitamin D [25(OH)D$_3$], also known as calcidiol.
2. The second occurs primarily in the kidneys, and forms the physiologically active 1,25-dihydroxyvitamin D [1,25 (OH) $_2$D$_3$], also known as calcitriol.

Vitamin D is actually a pro-hormone, and not really considered as a Vitamin. Regardless, it has been shown to be crucial in preventing cancer, and the mechanisms are fairly well understood. Vitamin D can no longer be thought *only* as a nutrient necessary for the prevention of rickets among children.

The supercharged 1,25-OH then can interact with the nuclear vitamin D receptor (VDR)-mediated control of target genes.[4] It is a transcription factor that plays an important role in calcium/phosphate homeostasis, cellular proliferation and differentiation, and immune response.[5] Vitamin D$_3$ also aids to control the development of a wide variety of serious diseases like breast cancer, osteopenia, multiple sclerosis, prostate cancers, etc. Vitamin D$_3$ should be considered as a pro-hormone and is vitally essential for overall health and human well-being.[6]

UV Light from the Sun: The Important Wavelength "Ultraviolet B"

There are two main wavelengths of ultraviolet light that come from the Sun: ultraviolet A ("UVA") and ultraviolet B ("UVB"). UVA is now considered to be "the harmful light" and UVB "the healthy light", since UVA can penetrate the skin more deeply and cause more free radical damage. UVB is essential to our health because it helps our skin trigger-off the chemical reactions that produces Vitamin D.

A Common Trend of Diseases in Countries Nearer the North and South Poles

Studies have shown that people who live in particular areas around the world where sunlight is lacking, the Northern and Southern poles and, indeed, to be more precise, in the Northern Hemisphere communities that are farther north, tend to suffer more from osteoporosis, type 1 diabetes, multiple sclerosis, rheumatoid arthritis, prostate cancer, colon cancer and breast cancer, apart from other diseases. It is extremely important for people living in these areas around the world to take their daily recommended doze of Vitamin D$_3$ supplementation.[7]

Is it Worth Buying Vitamin D supplements?

Most Vitamin D supplements are virtually worthless. This is because the vitamin D in milk and in most supplements is mostly in the form of Vitamin D$_2$ (ergocalciferol) and is synthetic. It is not the form of Vitamin D that you need to prevent cancer and other degenerative diseases. It is actually Vitamin D$_3$ (cholecalciferol) which is essential and is produced from the UVB rays of sunlight and can be considered as the "most affordable and absolutely free cancer-fighting nutrient in the world". If you do buy supplements, make sure you buy supplementations that are specifically labelled and contain "cholecalciferol or Vitamin D$_3$.[8]

The Sunscreen Protection Myth

Despite what we read and hear from sunscreen producers, sunlight is actually beneficial to us (especially the UVB rays of the Sun), and

sunscreens filter our UVB! The main chemical found in most sun-blockers or sunscreen lotions is Octyl Methoxycinnamate (OMC) which has been shown to kill mouse cells, even at low doses, and to be toxic when exposed to sunlight. Unsurprisingly enough, OMC is present in 90% of sunscreen brands![9]

The best formula for protection too long-term exposure to the rays of the Sun is an "Aloe Vera" natural gel or cream. However, whenever sunlight is not available (e.g. In the winter months) another excellent source of Vitamin D that can also provide beneficial supply of omega-3 fatty acids, DHA and EPA (which are protective against heart disease, cancer and other diseases) is Cod-Liver Oil (CLO).

Andrew Weil, M.D., has this to say about this topic:

> Low levels of vitamin D in the population as a whole suggest that most people need to take a vitamin D supplement. This may be especially true for seniors, as the ability to synthesize vitamin D in the skin declines with age.

There are numerous online websites where one can buy organic and natural Vitamin D_3 supplementations.

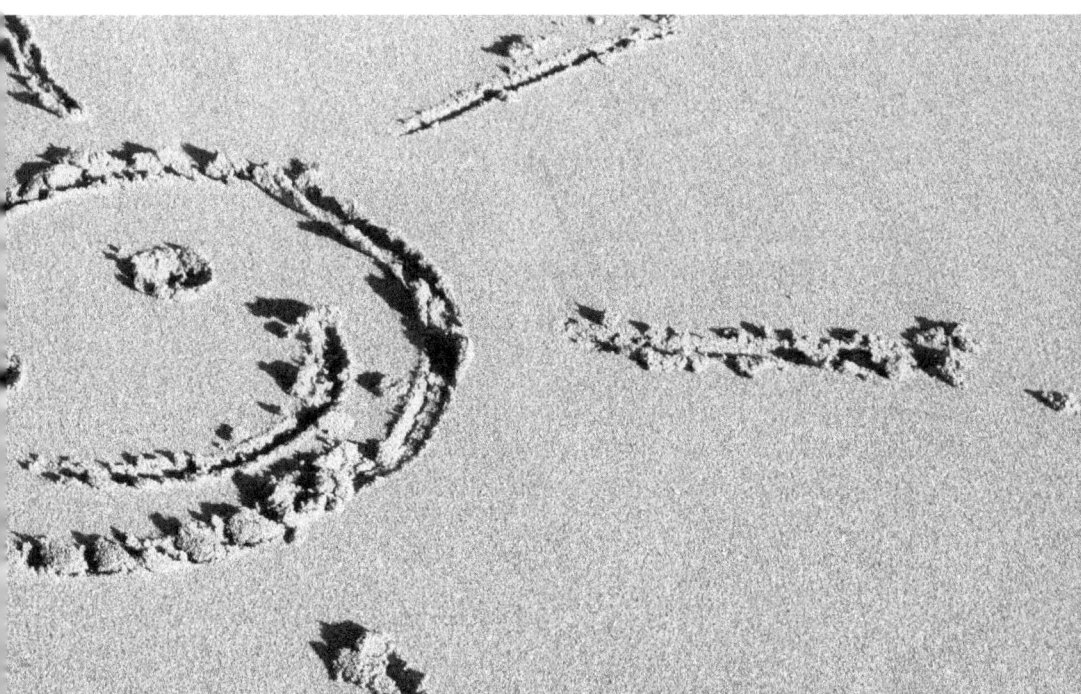

Caution must be exercised when taking CLO as one could overdose on Vitamin D. For this reason, CLO should be consumed only in the cool winter months. During the summer season, most people can get enough of this essential pro-hormone with 15-30 minutes of direct daily sunlight exposure, which is more than adequate to help start the synthesis process of Vitamin D from the precursor chemical located on our skin.[10]

Overall, fortified vitamin D foods, including fish and fish oils, are essential and are good sources of Vitamin D, but everybody should aim at, and select better, alternative foods to obtain adequate daily amounts of this vital pro-hormone. We must not forget that there is an unhealthy "price" attached to these foods: saturated fats, cholesterol, toxins, heavy metals (lead, cadmium, and mercury), hormones and antibiotics, heterocyclic amines and other carcinogenic agents, which are present whenever animal meat is eaten or is cooked at high temperatures.

Many people today still think that consuming fish or white meat (chicken, turkey or fish sources of foods) is healthier than eating so-called red meats (beef, pork, lamb, etc.). This is not true.

From studies carried out, it was found that for people who avoided red meat, but ate white meat regularly increased their risk of contracting colon cancer by up to 300%, when compared to people who ate no white meat. The same study also showed that those who ate beans, peas or lentils at least twice a week lowered the risk by 50% when compared to people who never ate such foods.

From the human heart perspective on health, the difference between "lighter" meats and red meats is insignificant. In comparison, it is like the difference between "regular" cigarettes vs. "light" cigarettes. All meats, from red meat to chicken, turkey, fish to liquid meat (dairy products) cause degenerative diseases. This is mainly because animal products contain high levels of sex hormones, saturated fat, cholesterol and animal protein, which, as we will see in later chapters, may increase the formation and development of disease.[11]

A survey of over 23,000 postmenopausal women has shown that fish consumption was positively associated with a higher incidence of breast cancer.[12] Several other study's confirmed these results. Eating fish increases the risk of cancer as well as the risk of it spreading to other parts of the body.[13] In addition to the high-fat levels in fish, one must keep in mind also that fish oils suppress the immune system.[14]

For safer Vitamin D alternatives, it is advisable to exercise outdoors where you can be exposed to direct sunlight (30 minutes per day), eat adequate amounts of cooked mushrooms, and/or take Vitamin D_3-labelled supplementations daily.

It has been estimated that 1,000 IU per day is the minimum amount needed to maintain adequate levels of vitamin D in the absence of sunshine, and that up to 4,000 IU per day of Vitamin D_3 supplements can be safely used with added benefits.[15]

The concentration of the functioning, hormonal form of vitamin D is tightly regulated in the blood by the kidneys. This active hormonal form of vitamin D also has *potent anti-cancer properties*.[16]

It has been discovered that various types of normal and cancerous tissues, including prostate cells, colon tissue, breast, ovarian and cervical tissue, pancreatic tissue and lung cancer cell lines all have the ability to convert the major circulating form of vitamin D, 25 (OH) D_3, into the active hormonal form, 1,25 (OH) $_2D_3$. Therefore, there is a local mechanism in many tissues of the body for converting the active form of vitamin D in the body. This is then elevated by sunshine exposure into a hormone that has anti-cancer activity. It could be that sunshine and vitamin D are protective factors for cancers of many organs that can convert 25 (OH) D_3 into 1,25 (OH) $_2D_3$.[17]

In the winter season and for those persons whom exposure themselves too little amounts of sunlight, especially within the Northern and Western regions around the world, it is healthy advised to add to their daily-dietary plan a beneficial supplemental form of vitamin D_3, also known as 1,25-dihydroxyvitamin D_3.

Notes for chapter 9

1. Holick, M.F. et al. (2004). Sunlight and Vitamin D for bone health and prevention of autoimmune diseases, cancers, and cardiovascular diseases. *The American Journal of Clinical Nutrition*, **80**(6), pp. 1678S-1688S.

2. Welsh, J., et al. (2003). Vitamin D_3 Receptor as a Target for Breast Cancer Prevention. *Journal of Nutrition*, **133**, pp. 2425S-2433S.

3. Holick, M.F., MacLaughlin, J.A., & Dobbelt, S.H. (1981). Regulation of cutaneous previtamin D_3 photosynthesis in man: skin pigment is not an essential regulator. *Science*, **211**, pp. 590-593.

4. Kato, S. (2000). The function of vitamin D receptor in vitamin D action. *J Biochem*, **127**(5), pp.717-722

5. Wang, Y, et al. (2012). Where is the vitamin D receptor? *Arch Biochem Biophys*, **523**(1). pp.123-33. doi: 10.1016/j.abb.2012.04.001.

6. Colston, K.W., et al. (1989). Possible role for vitamin D in controlling breast cancer cell proliferation. *Lancet* **1**, pp.188-191; Al-Qadreh, A. et al. (1996). Treatment of osteopenia in children with insulin-dependent diabetes mellitus: the effect of 1-alpha hydroxyvitamin D_3. *European Journal of Paediatrics*, **155**, pp.15-17; Nieves, J. et al. (1994). High prevalence of Vitamin D deficiency and reduced bone mass in multiple sclerosis. *Neurology*, **44**, pp.1687-1692; Peehl, D.M., et al. (2003). Pathways mediating the growth-inhibitory action of Vitamin D in prostate cancer. *Journal of Nutrition*, **133** (suppl), pp. 2361S -2469S.

7. P. Lips. (2007). "Vitamin D status and nutrition in Europe and Asia". *Journal of Steroid Biochemistry and Molecular Biology*, **103**(3-5), pp. 620–625. R. Andersen, et al. (2005). "Teenage girls and elderly women living in Northern Europe have low winter vitamin D status," *European Journal of Clinical Nutrition*, **59**(4), pp. 533–541.

8. Bollinger, Ty. (2006). *Cancer Step outside the Box* (5th ed). Chapter 12: Fantastic Foods & Super Supplements: Vitamin D, pp. 295-297. Published by Infinity 510^2 Partners.

9. Ibid.

10. Ibid.

11. Anderson, M. (2009). *Healing Cancer from Inside Out* Part 2: Reversing Cancer: High Sex Hormones, p. 69. Published by www.RaveDiet.com

12. Stripp, C., et al. (November 2003). Fish Intake Is Positively Associated with Breast Cancer Incidence Rate. *Journal of Nutrition*, **133**(3), pp.664-3669.

13. Young, M.R. (April 15th, 1989). Effects of fish oil and corn oil diets on prostaglandin-dependent and myelopoiesis-associated immune suppressor mechanisms of mice

bearing metastastic Lewis lung carcinoma tumours. *Cancer research,* **49**(8), pp. 1931-1936.

14   Griffin, P. (August 1st, 1998). Dietary omega-3 polyunsaturated fatty acids promote colon carcinoma metastasis in rat liver. *Cancer research,* **58**(15), pp. 3312-3319.

15   Holick, M.F. (2004). Vitamin D: importance in the prevention of cancers, type 1 diabetes, heart disease, and osteoporosis. *American Journal of Clinical Nutrition,* **79**, pp.362-371; Vieth, R., Kimball, S., Hu, A., & Walfish, P.G. (2004). Randomized comparison of the effects of the vitamin D$_3$ adequate intake versus 100 mcg (4000 IU) per day on biochemical responses and the wellbeing of patients. *Journal of Nutrition,* **3**(8) doi:10.1186/1475-2891-3-8

16   Donaldson. S.M. (October 20th 2004). Nutrition and Cancer. A review of the evidence for an anti-cancer diet. *Journal of Nutrition,* **3**(19): pp. 1-21.

17   Schwartz, G.G., *et al.* (1998). Human prostate cells synthesize 1, 25-dihydroxyvitamin D$_3$ from 25-hydroxyvitamin D$_3$. *Cancer Epidemiol Biomarkers Prev,* **7**(3), pp.91-395; Tangpricha, V., *et al.* (2001). 25-hydroxyvitamin D-1alpha-hydroxylase in normal and malignant colon tissue. *Lancet,* **357**, pp.1673-1674; Friedrich, M., Rafi, L., Mitschele, T., Tilgen, W., Schmidt, W., Reichrath, J. (2003). Analysis of the vitamin D system in cervical carcinomas, breast cancer and ovarian cancer. *Recent Results Cancer Res,* **164**, pp.239-246; Schwartz, G.G., *et al.* (2004). Pancreatic cancer cells express 25-hydroxyvitamin D-1alphahydroxylase and their proliferation is inhibited by the prohormone 25-hydroxyvitamin D$_3$. *Carcinogenesis,* **25**, pp.1015-1026; Mawer, E.B., *et al.* (1994). Constitutive synthesis of 1,25-dihydroxyvitamin D$_3$ by a human small cell lung cancer cell line. *Journal of Clinical Endocrinology Metabolism,* **79**, pp. 554-560.

10

Strategies to Help Improve Local Healthy-Dietary Lifestyle Changes

Each patient carries his own doctor inside him.

Norman Cousins
Anatomy of the Illness

On the Mediterranean island of Malta, where I currently reside, 1,152 deaths were reported due to diseases of the circulatory system in 2010, a decrease of 97 deaths from the year 2009. Cardiovascular disease is a leading cause of death, and accounts for 38% of all deaths.[1]

Diseases of the circulatory system, mainly ischemic heart disease, cerebrovascular disease and heart failure, rank as the most common causes of death. Acute and chronic lower respiratory tract infections were prominent causes of death in older persons. Diabetes mellitus is both a common cause of death as well as a significant risk factor in circulatory diseases. Lung, colorectal and breast cancer were the most prevalent causes of death due to malignancy in 2010.[2]

However, while the percentage of deaths due to cardiovascular disease in Malta has decreased over the past years, the percentage of deaths due to neoplasms (cancers) has increased. The number of deaths from neoplasm totalled 865, an increase of 13 deaths over 2009. According to the same 2010 Mortality Report, deaths in the mental and behavioral category were due mainly to dementia. Deaths in the endocrine category were, for the most part, due to diabetes mellitus.

While the increased awareness and incidence of diabetes is important, the altered nutritional habits of the Maltese population in recent decades have significantly affected the increasing levels of diabetes mellitus. Despite Malta's location, its diet is not in keeping with the accustomed Mediterranean diet. The Mediterranean diet includes more fish than red meat, more olive oil than butter, and lots of fresh fruits and vegetables (Townsend Rocchiccioli, 2005).

Malta's history of colonization has altered its cuisine, which currently offers a combination of tastes of many different cultures: diets that are higher in animal products (meat and dairy), fried foods, refined sugars, saturated fats and cholesterol. Regular consumption of these foods over time has led to increased rates of obesity in Malta, especially significantly increased rates of diabetes mellitus (Townsend Rocchiccioli, 2005).

Are Vegetable Oils Healthy, Including Olive Oil?

Some highly questionable "Mediterranean" diet studies[3] claim to show a "slower progression" of heart disease when people consume olive oil. Although, having a slower progression of heart disease hardly makes the case that olive oil is heart-healthy. Other studies have shown it is not olive oil at all, but the high fibre content in a traditional Mediterranean-type diet that accounts for lower rates of heart disease.[4] A diet is not heart-healthy unless it arrests and reverses heart disease and only a "whole-rich plant-based diet" which specifically excludes *all* oils – has been proven to do that. Adding oils to such a diet only impair its effectiveness.

The model for the Mediterranean diet harks back to the 1950s and before when people, especially on the island of Crete, were virtually free of heart disease – despite their indulgence of olive oil. This was not because of the olive oil, but because they ate primary whole grains and vegetables, a little fish and got lots of exercise. Today, their consumption of olive oil remains the same (if not increased), but their consumption of whole plant foods had plummeted, as has their exercise: the result, heart disease and obesity have skyrocketed.[5]

By using the brachial artery tourniquet test, it was found that olive oil constricted blood flow by a surprising 31 percent![6] Is this an important health issue? Yes! Because when arteries constrict, the endothelium (artery's lining) is injured, triggering plague build-up or atherosclerosis (heart disease).[7]

FACT BOX

Olive Oil is **100 percent fat**, contains high concentrations of saturated fat: the fibre, antioxidants, vitamins and minerals of the olive itself have been stripped away. In clinical studies on humans, olive oil has been shown to be as bad for the heart as eating roast beef. – JAMA, 1990, Vol. 263, pp. 1646-1652. Anderson, Mike. (2009). Healing Cancer from Inside Out, pages 96-100.

Olive oil has nothing to do with preventing heart disease and clinical studies have shown that it "actually" promotes arterial lesions.[8] "It may just be another magic bullet that is a big, fat fantasy promoted by the olive oil industry determined to get high levels of fat back into our diets."[9]

Leading health authorities in reversing heart disease with diet, including Dr. Joel Fuhrman, Dr. John McDougall, Dr. Caldwell Esselstyn and Dr. Dean Ornish, all agree that vegetable oils should be *excluded* from a heart-healthy diet. These are doctors who exactly reversed heart disease, as opposed to those who just talk about it or promote the greasy stuff; because of financial ties to the olive oil industry. In fact, there has not been a single case of heart disease arrest or reversal where any vegetable oil was a part of the diet. So, if olive oil is so "heart-healthy," why would doctors who reverse heart disease with diet purposely exclude it? Because they consider vegetable oils to be "heart dangerous" and so should you.[10]

The rates and trends of obesity and diabetes that affect these non-communicable diseases in the Maltese islands are some of the highest; when compared with those of other industrialized European countries. Prevention of obesity and diabetes is of paramount importance, particularly in childhood. An interdisciplinary public health campaign at all levels is vital, and should stress the importance of proper eating and lifestyle habits (Townsend Ricchiccioli, 2005).

No wonder that we have noticed an increase in cancer rates and other non-communicable diseases in our own backyard. We used to believe that our Mediterranean diet was once beneficial to our health. However, diseases have escalated – this could be mainly due to the dramatic unhealthy changes occurring in our daily-dietary habits, in-which the Maltese population have incorrectly adopted from westernized nations.

What can we do to Improve Health?

To start off, it is imperative to implement new policies and legislation in respect of food labelling. Why do we have labels on cigarette packets that say: "Smoking seriously harms you and others around you" but do not have other labels warning us about harmful refined or processed foods, including meat and dairy products? After all, foods that contain artificial colors, preservatives, refined sugars, cholesterol, and saturated fats can also damage our health. This is, in my opinion, a very valid observation.

Approaches to Health Promotion:
Examples of Healthy Eating and Lifestyle

We need to focus on improving our medical health-care system primarily focusing on behavioral, educational, empowerment, awareness and social change strategies that would lead to bettering health promotion: to improve healthy-lifestyle changes and implement eating habits. We should identify people who are at risk from disease; encourage individuals to take responsibility for their own health; increase knowledge and skills about healthy lifestyles; work together with clients or communities to meet their perceived needs; address class-based inequalities in health, race and gender, among other initiatives.[11]

Clinical scientists are now realizing that animal produce is harmful to people's health. People have the moral right to know which foods are **damaging** and which foods are truly **beneficial**. Food should be labelled according to legal regulations,[12] and any food item that is scientifically proven to contain unhealthy toxic chemicals should be clearly classified and labelled as harmful to public health.

We should consider spending more of our yearly health-care budget on advertising and promoting healthier foods for school children, like vegetables and fruits. We should not let our children eat junk food that can cause them long-term health problems later-on in life.

Another question: Why should the prices of many healthy foods be more expensive than unhealthy ones? We should subsidize local healthy foods and at the same time introduce or increase taxes on unhealthy food products: thus, this may help encourage people to buy and consume wholesome plant-food at a lower cost.

We should introduce school dinners in schools and universities: where students will be welcomed with nutritional fruits and vegetables, whole-grains of bread, pasta and rice, including nuts, seeds, legumes and other healthy food choices on their menu.

One other thing we could do is to explain to children why consuming animal-based foods and eating less healthy plant foods can lead to chronic degenerative diseases later on in their life; including the impact animal produces has on global warming and about animal welfare issues. It is difficult to convince people to make a complete transition to a plant-based diet, but there can always be a start. We must remember that the more people eat healthy plant foods, the less likely it will be that they will eat unwholesome ones.

This may loom like a very big challenge, but the long-term effects will be worthwhile – as all this – would be for the benefit of our children's future health. It is important that we remember – if we have healthy children, we shall surely build a strong, productive and healthy nation.

This is not scaring children or parents, but rightfully educating them and empowering them to fully fathom the importance of choosing and consuming more plant foods, rather than animal foods.

These are only some of the critical strategies that our governments can endorse to accomplish a truly superior health-care system.

We need to embark on academic approaches towards healthier food choices, so that parents can choose better and more sustainable foods for their children and, at the same time, look after their own health.

One final proposal is to increase financial support, using more of our yearly budget funds to publicize and educate people about disease prevention: using leaflets, TV, newspaper advertisements, etc. about family dietary planning incorporated with lifestyle medicine. These tactics will most certainly lead to less medical visits, less hospitalization, less buying of pharmaceutical drugs and, more significantly, fewer incidences of chronic-related diseases.

Drugs are life-savers in the treatment of acute diseases, especially infectious diseases, during allergic reactions, in reducing inflammation, traumas and in the relief of pain, but for the prevention and reversal of chronic diseases, especially cancer, a new dietary/lifestyle approach is desperately needed.

Plant Anti-Cancerous and Anti-Angiogenic Medications

Drugs and other medications for the prevention and reversal of cancer *should not* be found in a form of a liquid, pill or capsule, but in their natural shape and state: an organic anti-cancerous and anti-angiogenic form of plant food.

Differences between Conventional and Lifestyle medicine

Conventional	Lifestyle
Treats individual risk factors	Treats lifestyle causes
Patient is often passive recipient of care	Patient is an active partner in care
Patient is not required to make big changes	Patient is required to make big changes
Treatment is often short term	Treatment is always long term
Responsibility falls mostly on the clinician	Responsibility falls mostly on the patient
Medication is often the "end" treatment lifestyle change	Medication may be needed but as an adjunct to
Emphasis is on diagnosis and prescription	Emphasis is on motivation and compliance
Goal is disease management prevention	Goal is primary, secondary and tertiary disease
Little consideration of the environment	Consideration of the environment
Little consideration of animal welfare	Consideration of animal welfare
Side effects are balanced by the benefits	Side effects are seen as part of the outcome
Referral to other medical specialties	Referral to allied health professionals as well
Doctor generally operates independently on a one-to-one basis	Doctor is coordinator of a team of health professionals

According to T. Tarver,

> Cures for cardiovascular disease, cancer, diabetes, and obesity have eluded scientists for decades, but research in nutritional genomics suggests that halting the progression of these diseases may be as simple as a dietary intervention.[13]

When the treatment for a chronic condition takes the form of a drug this will only mask the problem and will "only" manage a patient's symptoms. People want to know "how" and "why" did their symptoms transpire in the first place. They want to know what is the "real underlying cause" of their health condition. In many cases the cause could have been due to that person's bad diet and unhealthy lifestyle.

We should have no more medications to relieve chronic diseases, no more operation procedure for heart diseases, cancers, etc. On the contrary, we should teach people how to look after themselves with excellent, natural, nutritious plant foods and lifestyle medicine.

According to Dr. M.A. Hayman,

> "Only sincere political commitment will be effective in the prevention of disease".

This new government health plan could save our own government millions of euros per annum in hospital and drug fees. Even so, more importantly, it may also save the lives of thousands of patients with chronic-related diseases. There is only one-way forward, and that is to educate the general public, including those who work in the medical profession, about the important health implications of *"Lifestyle Medicine"*.[14]

What is Lifestyle Medicine?

Lifestyle Medicine is the use of lifestyle interventions in the treatment and management of disease. Such interventions include:

- Diet (Nutrition)
- Stress Management

- Smoking Cessation
- Exercise
- Other interventions using non-drug modalities.

An emerging, growing body of scientific evidence has demonstrated that lifestyle intervention is an essential and critical component in the treatment of chronic diseases – that can be as effective as medication, but without the risks and without any unwanted side-effects.

Lifestyle Medicine, without any doubt, is the "Primary Disease-Preventative Management Treatment", that can lead to the gradual eradication of non-communicable diseases from today's societies: towards the reduction of obesity, diabetes, heart disease, strokes, cancer and other diseases, in which have progressively increased within the Maltese population.

Notes for chapter 10

[1] World Health Organization (WHO). (2005). *Regional Office for Europe: Highlights on Health in Malta.* Retreived from www.euro.who.int/document/e72500.PDF

[2] Department of Health Information and Research: Maltese National Mortality Registry Annual Report (2010). The International Statistical Classification of Diseases and Relative Health problems (ICD-10) pp.19-20; Townsend Rocchiccioli, J., et al. (March, 2005). Diabetes in Malta: Current findings and Future Trends. *Malta Medical Journal.* Volume 17.

3   For example, Lyon Diet Heart Study: Note that after nearly four years into the study, 25 percent of the subjects following the "heart-healthy" Mediterranean diet had either died or experienced some new cardiovascular event.

4   European Journal of Clinical Nutrition (2002). **56**, pp.715-722.

5   The same sort of argument is made by those peddling any oil. Coconut oil vendors, for example, will argue that despite its very high saturated fat content, populations consuming large quantities of coconut products (such as the Philippines), have low rates of heart disease. Again, this is due to their overall (traditional) low fat, low cholesterol diet. The low rates of heart disease was achieved not because of the coconut oil, buy despite it.

6   Vogel RA, *et al.* (Nov 1st 2000). The postprandial effect of components of the Mediterranean diet on endothelial function. *Journal of the American Collage of Cardiology,* **36**(5), pp.1455-1460.

7   High fat meals block the endothelium's ability to produce nitric oxide, which is a vasodilator and critical to preserving the tone and health of blood vessels. Moveover, fat should be released in the blood stream slowly and you can only get that slow release by eating natural plant foods in their natural packages that still contain fibre. Anderson "Healing cancer from inside out", 2009 pp.98.

8   Blankenhom, DH *et al.* (1990). The Influence of Diet on the Appearance of New Lesions in Human Coronary Arteries. *JAMA,* **263**(12), pp. 1646-1652.

9   Anderson, M. (2009). Healing Cancer from Inside Out. Chapter 3: The RAVE Diet & Lifestyle, pp. 96-100. Published by www.RaveDiet.com

10 Anderson, "Healing Cancer from Inside Out", 2009.

11 Naidoo, J. & Wills. (2000). *Health Promotion: Foundations for Practice (2nd ed)*, p. 102. Published by Bailliere Tindall.

12 Nutritional Labelling for Foodstuffs Regulations. (1998), L.N. 247 of 2008.

13 Tarver, T. (October, 2012). *The Chronic Disease Food Remedy: Feeding the minds that feed the world.* **66**(10). Retreived from http://www.ift.org/Food-Technology/Past-Issues/2012/October/Features/The-Chronic-Disease-Food-Remedy.aspx

14 Hayman, M.A. *et al.* (November/December, 2009). Lifestyle Medicine: Treating the Cause of Disease. *Alternative Therapies,* **15**(6), pp. 12-14; Ornish, D. *et al.* (1990). Can lifestyle changes reverse coronary heart disease? The Lifestyle Heart Trial. *The Lancet,* **336**, pp.129-133; Greger, M. (Novemver 4th, 2013). *Lifestyle medicine: Treating the cause of disease.* Volume 15 video. Retreived from http://nutritionfacts.org/video/lifestyle-medicine-treating-the-causes-of-disease/

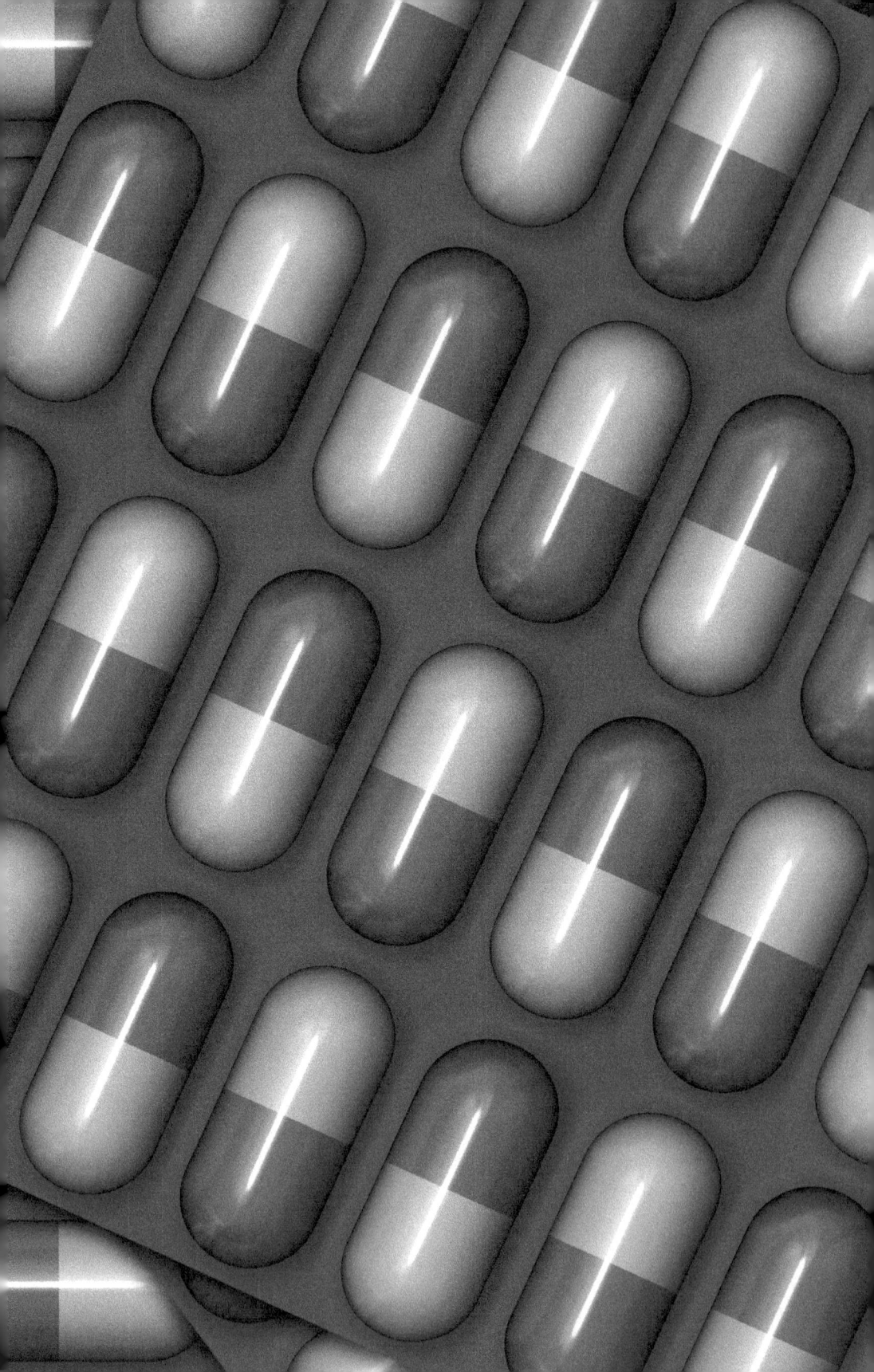

11

The American Medical and Pharmaceutical Cartel

Modern allopathic medicine is the only major science, stuck in the pre-Einstein era.

Charlotte Gerson
*Healing the Gerson Way:
Defeating Cancer and other Chronic Diseases*

If I were to mention all the environmental toxins and other detrimental chemicals found in nature, and within the foods we eat; that are harmful to human health, this book will be too time-consuming and may never be completed. Nevertheless, in later chapters, I *will* mention some of the most significant toxins and chemicals that you may have heard or know of, which have been proven beyond reasonable doubt: documented as causing biological and metabolic harm to the human body and to other experimental mammals.

Initially, however, it is important that we look back in time when the problem first started. The time when the industrial revolution was in full swing in an attempt to balance socioeconomic problems, especially unemployment. We also need to consider how the drug and pharmaceutical companies are controlled by extremely powerful organizations, whose power is so great that no large, billion-dollar, Food and Drug Company in America can say no to "The Rockefeller, Carnegie and Ford Foundations".

The Policies on Drugs and of the Medical Profession in the USA

"They (the Rockefellers) feared the temptation of wealth, but once a visitor described their home estate as a kind of place God would have built, if only he had the money."[1]

They amassed a fortune that outraged a democratic nation, then, they gave it away, in a way that reshaped America. They were the closest thing the country could have had to a royal family, but the Rockefellers shunned the public eye, and retreated behind the walls of their home in New York.

The family found themselves hounded by the controversy surrounding John Davison Rockefeller, "King of Standard Oil." Vilified as a ruthless predator, as evil as Cain, he had created an industrial empire, and a personal fortune on a scale the world had never known before.

There is an old saying: "He who pays the piper, calls the tune". This is one of those eternal truths that exist – and always will exist – in business, in politics, and in education.

Yes, he who pays the piper does call the tune. It may not be possible for those who finance the medical schools to determine what is taught in every minute detail. Yet, this is not even necessary to achieve the cartel's desired goals.

We need to keep in mind the low level of health education in the U.S. prior to 1910, the importance of the Flexner Report in dramatizing the need for reform, the role played by the Rockefeller, Carnegie and Ford Foundations in implementing the said Flexner Report, and the use of foundation funding as a means to gain control over American medical schools.

One can be sure, however, that there is total control over what is *not* taught, and that under no circumstance will even one of the Rockefellers' shiny dimes ever go to a medical college, a hospital, a teaching staff, or a researcher who may hold the "unorthodox" view that the best medicine is, after all, in *nature*.

Because of its generous patron, orthodoxy will always fiddle a tune of synthetic drugs. Whatever basic nutrition will be allowed into the melody, it will definitely be minimal, at best, and it will be played repeatedly: that natural sources of vitamins are in no way superior to those that are manufactured or synthesized. The day when orthodox medicine finally embraces the field of nutrition will be the day when the cartel behind it has also monopolized the vitamin and food product industries that are essential to it – and not one day before.

What Do Doctors Know About "Basic Nutrition"?

Meanwhile, doctors are forced to spend hundreds of hours studying the names and effects of all kinds of man-made drugs. They will be lucky if they will ever receive even a portion of a single course on basic nutrition. Many receive none at all. The result is that the *average doctor's wife or secretary may know more about practical nutrition than he does.*[2]

On average, out of thousands of hours of pre-clinical instruction concerning important diagnostic procedures, metabolic clinical diseases, and pharmaceutical drugs, American doctors *merely* get around 24 hours on nutrition, and most only get between 11 to 20 hours.[3]

However, we find that the cartel's influence over the field of orthodox medicine is felt far beyond medical schools. After the doctor has struggled his way through many years of studying, which is what the cartels have decided is best for him to learn, he then goes out into the world of medical practice and is immediately embraced by the other arm of cartel control: The American Medical Association.

The American Medical Association (AMA) is the union for the medical establishment. But truth, workability, effectiveness and health are not what it is concerned with. The AMA always reports handsome profits whenever the drug companies prosper. A large percentage of the income of the AMA is derived from drug companies' advertisements on its journal.

The AMA has acted ruthlessly to destroy and silence alternative therapies and practices which threaten the traditional medical establishment. It is all about power and money, and to hell with the health of the public. This has been its *modus operandi* (method of operation) since its inception.[4]

There are numerous relationships and connections between the drug companies, both national and international. These include, IG Farben, a mammoth German drug company with numerous cartel agreements throughout the world; the Food and Drug Administration (FDA); the American Medical Association (AMA); the American Psychiatric Association (APA); various "professional" journals and publications; research companies and organizations; philanthropic foundations; government organizations like the National Institute of Mental Health (NIMH); international organizations like the World Health Organization (WHO), and the medical schools.[5]

How the Rockefellers Control the F.D.A. and the Cancer Industry

"An unholy alliance formed between the American Medical Association, the F.D.A., and the Rockefeller Foundation."[6]

To ensure compliance from medical institutions, the Rockefeller Foundation frequently insisted that medical schools place Rockefeller employees on their board of directors. The new legally enforceable medical monopoly paralleled the past Rockefeller monopoly of the petrochemical industry of previous times. Instead of owning all petroleum, the Rockefeller Empire now controls virtually all medicine.

Rockefeller owned the lion's share of the chemical industry, which would later be called the "Pharmaceutical" industry. During his life, John D. Rockefeller Sr, refused to take his own medicine. Instead, he used traditional holistic medicines for his health. It was after the hijack of the medical schools that the true carnage of polio, heart disease and cancer exploded.

G. Edward Griffin, in his book, "World Without Cancer", has made explicit the fundamental, systematic wrongdoings, which has emerged out of the various crosscurrents that make up the F.D.A. He showed how underpaid civil servants "play it safe"; how drug companies and their Washington lawyers put unending pressure on bureaucrats; how academic medicos control the approval process and restrict the individual doctor's choice; the revolving door employment between F.D.A. and universities/drug companies and the behind-the-scenes political deals. According to Griffin, the F.D.A. did two things:

1. They "protected" the big drug companies, for which they were subsequently rewarded.
2. Using the government's police powers, they attacked those who threatened the big drug companies, even start-up companies with new products and break-through miracle drugs such as DMSO, or natural health store products such as food, vitamins, minerals or other self-healing (non-drug, non-doctor) methods.

What Griffin said about the F.D.A.

First, it (F.D.A.) is providing a means whereby key individuals on its payroll are able to obtain both power and wealth through granting special favors to certain politically influential groups that are subject to its regulation. This activity is similar to the "protection racket" of organized crime whereby for a price one can induce F.D.A. administrators to provide "protection" from the F.D.A. itself.

Secondly, as a result of this political favoritism, the F.D.A. has become a primary factor in that formula whereby cartel-oriented companies in the food and drug industry are able to use the police powers of government to harass or destroy their free-market competitors.

And thirdly, the F.D.A. occasionally does some genuine public good with whatever energies it has left over after serving the vested political and commercial interest of its first two activities.[7]

Griffin concluded by saying that: "There is only one solution. No reform will work. *Repeat: no reform will work.* No changing of personnel will have any long-term effect. No new laws dealing with regulations".[8]

There is only one solution... and it was provided by a Southern doctor now living in New York City who has observed the monster in action for many years. A certain Raymond Keith Brown, M.D., outlined the solution in his book, "AIDS, Cancer, and the Medical Establishment".[9] He described how the power which F.D.A. has to approve drugs and technology has to be replaced with the solitary role of testing for effectiveness and safety, the results being the basis for F.D.A. labelling. The individual physician and individual patient would regain the responsibility to use or not to use a given drug or technology.

Dr. R.K. Brown's Recommendations

The FDA should follow a simple rating system for effectiveness and safety. Effectiveness would fall into one of three categories: effectiveness unconditionally proved, effectiveness conditionally demonstrated and effectiveness undetermined.

Safety could also be categorized in the same manner, and the appropriate designation, then affixed to all products or containers. The judgement of the effectiveness of any medical product or device should not be vested in any governmental agency or institution, but should be returned to the province of the individual physician.

> Freedom of choice for medical materials, therapy and methods must be put on the same footing as civil liberties and as vigorously protected.

One of the best scholars in this field, Robert G. Houston, says simply: "There should be curbs on the F.D.A., on its powers to intrude into the private practice of medicine (...) the F.D.A. should not be dictating to doctors what they can and cannot do".

Richard Ericson, a dedicated husband of a cancer victim, eloquently concurred:

> A physician should be able to prescribe any type of cancer treatment that he considers best for the patient, with the patient's consent and knowledge, without stringent governmental regulations that are now in force. Congress should consider such problems when new guidelines are enacted. Only when F.D.A. concentrates on the blatant health menaces, such as overtly misleading health product claims or drugs shown to cause death and injury. Only when F.D.A. ceases to be the bully boy for the big drug companies and other vested interests, and only when F.D.A. again allows physicians, non-conventional healers and their patients some choice of therapeutic treatments... will it regain its legitimate government function. In its present form, it is like a malignant beast harming society rather than serving it.[10]

What about The American Cancer Society? Guess when this organization was founded and by whom? It was founded by John D. Rockefeller Jr. in 1913 during a "donation" ceremony at Harvard University. Do you really believe that they will someday announce a cure for cancer and cut themselves off from truckloads of money?

According to the U.S. census, cancer is the second biggest killer in the United States, but most of these people die from the treatments, not from the cancers. Pharmaceutical drugs are the fourth biggest killer. The U.S. has the most dangerous, yet the most expensive, medical system in the world. It causes 60% of all U.S. bankruptcies.[11]

The common factor in each is the presence of representatives of the "drug" approach to handling "illness". The drug companies are at the "top" of the hierarchy, and they are the ones who make the decisions. Why? It is because they are the source of the funding in all cases.

He who pays the piper calls the tune. This is difficult for most people to accept because "modern medicine" is a major social institution which, naturally, we all assume to be inspired by the human values of honesty, decency, and a genuine intention to help.

Sadly, this is just not the way it is. And it has not been that way for at least 100 years! There is corruption in many other aspects of life. Governments earn more money during wars by manufacturing and selling machinery, guns, armory and other weaponry devices. These unholy and evil tactics outnumber the money profile cartels made to compensate for the deaths and lives of millions of people worldwide.

But what does usually happen when people get unwell? The sickly and poor, the uneducated, and such like to rely on and take their so-called "good" doctor's advice, who, of course, have received almost no training at all on human nutrition. The doctor advises and prescribes to them what he thinks is the best medical prescription for them – no doubt with the intention to control their symptoms of their medical condition – but in most cases what happens is that the prescribed drugs will only make their condition worse.

The drug industry is constantly trying to convince us that drugs are good for our health, while nutritional supplements and healthy foods are somehow bad for us. This same line of nonsense is also repeated by the F.D.A. which goes out of its way to censor the truth about the healing properties of natural foods like apricot seeds, walnuts, cherries and berries.[12]

The drug industry, including the F.D.A., is, of course, just plain wrong about all this. Although their advertisements show happy, healthy people consuming pharmaceuticals, in the real-world, people who take such pharmaceutical drugs are extremely unhealthy, depressed and highly toxic.[13]

Health promotion, education and food awareness is extremely vital in order to effectively open up people's minds about healthy food choices and how they can improve their health and prolong their life.

We must be aware that various governments, health organizations, large corporations, the pharmaceutical industries and certain food establishments, through their policies, schemes and tactics, have found a brilliant way how to brainwash most people's minds. The majority, unfortunately; do not know the truth, or do not understand it, or do not make the effort, or do not have the time to learn the truth behind this scam, these evil doings, that some governments are hiding from them. The general population should be advised on the truth of the matter of the politics and the power of these food and drug industries.

We must not confide in these mighty powerhouses of world politics, on whom we think we should rely on; just because we assume they hold the truth about our health, medicine, power and economics. We need to open our minds, even our government's mind, and other people's minds, who might not actually be aware of what is really happening within or outside their own doorsteps.

The Rockefeller's plan was a smashing success, and conflicts of interest between Big Pharma and Big Medicine continue to this day. In his book, "Cancer-Gate: How to Win the Losing Cancer War",[14] Dr. Samuel Epstein demonstrates that over the past century, the National Cancer Institute (NCI), the American Society of Criminology (ASC), and the AMA have all become corroded with major institutional and personal conflicts of interest with Big Pharma.

It is a simple economic equation: keeping the public ignorant of the causes of cancer results in most cancer patients, more cancer incidences that result in more sales of chemotherapy drugs, radiation and more surgery.

You see, money rather than moral ethics, is the deciding factor for the cancer industry and the Medical Mafia. To be honest, their goal is to provide *temporary* relief by treating the symptoms of cancer with drugs, while never addressing the *cause* of the disease. This ensures regular visits to the doctor's office and requires the patient to routinely return to the pharmacy to refill his or her prescriptions. This is what the game is all about: plain and simple. It is a shame that many, westernized countries follow the immoral ways of the American health care system. For these very reasons, there are millions of unnecessary cancer deaths worldwide due to these political, unethical and immoral practices.[15]

The American health care system is one of the worst health care systems in the world; hence countries that follow their ways of medicine, treatments and therapies, will not succeed towards the prevention, control and cure of any chronic-pathological human disease.

We must find our own way to combat diseases, and not follow the erroneous ways and awful examples of others, whom we and our governments consider to be concerned about human health, but who in reality have a different objective, and for whom the drug cartel and money laundering is more important than saving human lives. We must focus more on health prevention (diet and nutrition, incorporating healthier lifestyle changes) if we want to succeed where other so-called "superior" countries' health care systems have failed. Diet and lifestyle medicine should be the primary focus on our own health care system agenda, especially concerning disease prevention.

Notes for chapter 11

1. American Experience, PBS. Article about the film, The Building of Rockefeller Centre. http://www.pbs.org/wgbh/americanexperience/features/general-article/rockefellers-center/?flavour=mobile

2. Parker, W.A., et al. (August 11th, 2011). They think they know but do they? Misalignment of perceptions of lifestyle modification knowledge among health professionals. *Public Health Nutrition,* **14**(8), pp.1429-1438.

3. Adams, K.M., et al. (April 2006). Status on nutrition in medical schools. *The American Journal of Clinical Nutrition,* **83**(4), pp. 941S-944S.

4. Griffin, G.E. (1974). *World without Cancer: the Story of Vitamin B$_{17}$*. (3rd ed) Published by American Media.

5. Ibid.

6. Griffin "World without Cancer", 1974.

7. Ibid.

8. Ibid.

9. Brown, R.K. (April 1st, 1986). *AIDS, Cancer and the Medical Establishment.* (1st ed). Published by Robert Speller & Sons.

10. Ericson, R. (November 1st 1979). *Cancer Treatment: Why So Many Failures? What you can do about it?* Published by GEPS Cancer Memorial.

11. Article was posted online and taken from "Birth of a New Earth" on March 4th, 2012. Barcelo, J. http://vimeo.com/99425497 video. The Rockefellers, the FDA and the Cancer Industry. http://birthofanewearth.blogspot.com/2012/03/how-rockefellers-control-fda-and-cancer.html

12. Adams, M. (2010). NATURAL NEWS: The Drug Industry: Retreived from http://www.naturalnews.com/028789_pharmacies_health_food_stores.html

13. Ibid.

14. Epstein, S.S. (February 28th 2005). *Cancer-Gate: How to Win the Losing Cancer War: Policy, Politics, Health and Medicine.* (1st ed).Vicente Navarro Series. Published by Baywood Pub Co.

15. Bollinger, Ty. (2006), *Cancer: Step outside the Box.* (2006). Published by Infinity 510 Squared Partners.

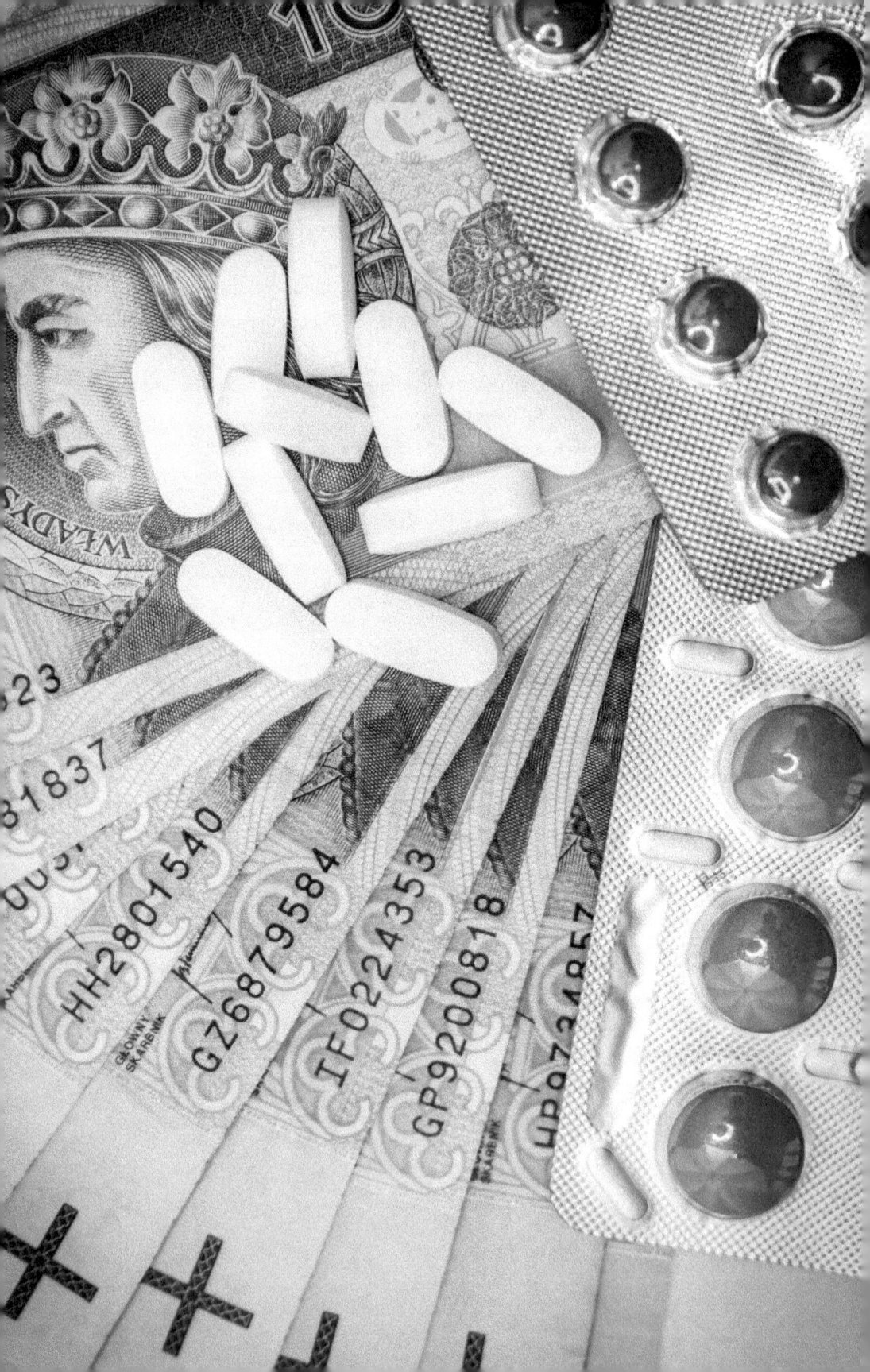

12

The War on Cancer

We are not dealing with a scientific problem. We are dealing with a political issue.

Samuel Epstein, M.D.

We have a multi-billion dollar industry that is killing people, right and left, just for financial gain. Their idea of research is to see whether two doses of this poison is better than three doses of that poison.

Glen Warner, M.D., oncologist

The Billion-Dollar Cancer Industry: The Blinded Truth Revealed

What you have already read in this book most probably has challenged everything you have heard since you were born. From the crib, we have been effectively taught to believe everything we read in newspapers, health articles, on the Internet, what we hear on the radio, and what we watch on TV about health matters.

Large pharmaceutical companies, the cancer industry, and many other food industries focus only on selling their so-called "health products" to make huge profits – but the health of the general public in most cases plays second fiddle. Primary disease intervention, that is, what we should focus on, is avoided. For this very reason, cancer and other chronic-related diseases have prevailed.

Ty Bollinger from his book, *"Cancer Step Outside the Box"*, thus ascertains the situation:

> The supposed "War on Cancer" is little more than a grand illusion conjured up by the cancer establishment's propaganda gurus. The formula is eons old. Repeatedly chisel your message into people's psyches: "cancer breakthrough", "scientists say they are turning the tide on cancer". We become unwitting human satellites, bouncing the deception from one person to another.[1]

There never was a determined, no-holds-barred war on cancer. There is a fanatical and hate-filled war being waged against the few courageous doctors and innovative healers who prescribe natural treatments. There is a war of protectionism: protecting the status quo, protecting the grant money trough and, above all, protecting the pharmaceutical cartel's monopoly.[2]

There have been at least a dozen very promising cancer treatments in the last seventy years: the Rife frequency machine,[3] Laetrile,[4] Hoxsey, Antineoplastons, Coley's Toxins, Glyoxylide, Hydrazine Sulfate, Krebiozen, Immuno-Augmentative Therapy, Brandt/Kehr Grape Cure, Essiac Tea protocol, the healthy use of selenium, high doses of intravenous Vitamin C, Insulin Potentiation Therapy, sodium bicarbonate and the Gerson therapy. These so-called "natural cancer treatments", all have 3 things in common:[5]

1. The people advocating the therapy are branded charlatans or quacks.
2. No one has an interest in preventing cancer because dietary prevention does not produce big pay days.[6]
3. These natural, non-harmful treatments have been denounced as worthless by most medical scientists, the medical establishments, or those who have been selling them out for generations.[7]

In 1900, the risk of cancer was 1 in 30. In 1980, it rose 1 in 5. In 1995, it became 1 in 3. Today, after billions of dollars in research, your chance of dying from cancer is greater than 1 in 2. As Ralph Moss pointed out, if the process continues at this rate, cancer deaths in the U.S. should be eliminated by the year 3508 – just a little more than 1,500 years from now![8]

Whenever a doctor tells a sick patient that there is no cure, what he basically means is that there are no drugs that can control the symptoms of the disease. It does not mean that the person *cannot* be healed. For example, there is a cure for heart disease. Clinical scientific evidence has shown that a person with heart disease can live a normal life, without conventional treatments, and that the production of plaque formation leading to the build-up of atherosclerosis in a person's arteries can be reversed by the body with a change in diet and lifestyle. Intervention: intensive lifestyle changes (10% fat whole foods vegetarian diet, aerobic exercise, stress management training, smoking cessation and group psychosocial support for 5 years).[9]

In other words, if he wants to be healed the patient must change his behavior and take charge of his or her own health.

Note that the "body as machine" paradigm works very well in many cases. If you break an arm or leg, the doctor will work with that part of the machine and repair its parts. If you are wounded by a gun-shot bullet, the doctor mechanically extracts the bullet, fixes the damaged area, and the body once again slowly heals.

The problem with the mechanical approach is that it does not work at all with systemic chronic-failure diseases – diseases that manifest themselves as a result of a systemic failure or degeneration of the body. Yet, systemic diseases have symptoms, but treating the symptom is not going to correct the underlying systemic cause of the problem. The marvelous and miraculous effect of dietary treatments is that they treat the body as a whole and work to re-balance a body that has become seriously out of equilibrium and homeostasis. This is what happens with *all* our chronic diseases.[10]

Although, some breast cancer studies, like "The Women's Healthy Eating and Living (WHEL), randomized a trial of more than 3,000 women with breast cancer.[11] The final results concluded that there were no health benefits of eating more fruits and vegetables and less fat. Headlines across the news read: "No Cancer Benefit Found in Mega-Veggie-Diet Study", "Dietary Hopes Dashed for Breast Cancer Patients", "Healthiest Diet Made Little Difference to Breast Cancer Survivors", and so on. But, what the Journal of the American Medical Association (JAMA) and the newspapers failed to mention was that the study was, in fact, fraudulent.[12]

The women in the study simply lied about what they ate (which is not uncommon in survey studies). How do we know they were lying? Because their food diaries indicated that the women, who were obese, were decreasing their average daily caloric intake by 181 calories per year over six years. That would have resulted in at least a 12-pound loss of weight – and yet they actually gained weight during the study! There can only be two conclusions:

1. Either they violated the laws of physics, or
2. They lied about what they were eating.

On top of that, their increase in fruit and vegetable consumption was extremely minor (small) and the authors resorted to using relative (dishonest) numbers in order to exaggerate the increase in fruit and vegetable consumption, as well as the decrease in fat consumption. When put into absolute numbers, the differences were minimal, at best, and one would hardly call it an anti-cancer diet. In fact, these poor women were eating the standard American diet, which causes cancer!

The fraud that was committed by the authors was to tell the public that cutting fat and increasing plant food consumption does not cause a change with cancer. Yet, there are mountains of clinical documenting scientific evidence telling us otherwise.[13]

The real tragedy is that through the news headlines the entire nation became aware of the results of this study (that fruits and vegetables do not fight cancer) but only a few thousand people know about its fraudulent methods. So, it is important to keep your scepticism intact, and keep our eyes open about certain so-called "scientific studies".[14]

A radical change in cancer research and treatment is desperately needed. The natural, nutritional and other innovative approaches should be studied and made available to cancer patients immediately. Most importantly, we must have medical freedom of choice. For us to achieve these changes, we have to overcome a far tougher opponent than cancer: a true, unwinnable battle against the $110 billion-a-year cancer industry. Ultimately, our greatest enemy is apathy that will cause pain or death for chronically-ill patients, who have not been given the hidden information necessary to reverse their own disease.

They have been misinformed or lied to, and the vital information and the real cause hidden away: just like many other common chronic degenerating diseases inflicting today's societies. It makes one sad to think this, but it is the truth of the matter: people are dying and suffering from many diseases that could have been prevented if the true information had not been hidden from and accepted by the general public and the medical care industry.

Cancer patients are, sadly, suffering or dying with cancer: which indeed, could actually have been prevented, controlled, reversed and, in some cases, cured by just changing their dietary choices and lifestyles.

To succeed, we need people who are willing to stand up to this cause and speak out, undeterred by being labelled "politically incorrect". We need people with plenty of good old-fashioned guts, character, and an iron will to see it through until the job is done. But we will prevail. It is inevitable. It is because when good men and women put their minds to something, the mightiest walls of oppression can and will be shattered.[15]

I hope that *you* will be one of these persons.

Notes for chapter 12

1   Bollinger, Ty. (July 17th, 2006). *Cancer: Step outside the Box*. Published by Infinity 510 Squared Partners.

2   Ibid.

3   Lynes, B. (July 1st, 1987). *The Cancer Cure That Worked: 50 years of Suppression*. Published by BioMed Publishing Group.

4   Day, P. (May 1999). *Cancer: Why we're still dying to know the truth*. Credence Publications.

5   Healing Cancer Naturally: History of Alternative Treatments (2): *Why frequently successful cancer treatment approaches such as the Royal Raymond Rife's and others are not being offered to cancer patients today*. http://www.healingcancernaturally.com/medical-history-2.html.

6   Anderson, M. (2009). H*ealing Cancer from Inside Out. Part 1: Say Goodbye to the Cancer Industry*, pp.43-47. Published by www.RaveDiet.com

7   Walter, R. (August 1st, 1992). *Options: the Alternative Cancer Therapy Book*. (1st ed). Published by Avery Trade.

8   Moss, R. (2006). The Moss Reports Newspaper. Retrieved from http://cancerdecisions.com/ . Bollinger, "Cancer: Step outside the box", 2006.

9   Ornish, D., et al. (2008). *Intensive Lifestyle Changes for Reversal of Coronary Heart Disease*. JAMA, **280**(23), pp.2001-2007. Ornish, D., Brown, S.E., Scherwitz, L.W., *et al.* (1990). Can lifestyle changes reverse coronary atherosclerosis? The Lifestyle Heart Trial. *Lancet*, **336**, pp.129-133.

10 Ornish, "*Intensive Lifestyle Changes for Reversal of Coronary Heart Disease*", 2008.

11 Pierce, J.P., *et al.* (2007). Influence of a Diet Very High in Vegetables, Fruit, and Fibre and Low in Fat on Prognosis Following Treatment for Breast Cancer. *JAMA*, **298**, pp. 289-298.

12 For an excellent analysis of this study, see The McDougall Newsletter, July 2007, Vol.6, No.7.

13 Kwan, M.L., *et al.* (2009). Dietary Patterns and Breast Cancer Recurrence and Survival among Women with Early-Stage Breast Cancer. *Journal of Clinical Oncology*, **27**(6), pp.919-926. doi: 10.1200/JCO.2008.19.4035.

14 Bollinger, "*Cancer: Step outside the Box*", 2006.

15 Ibid.

13

Medical Compliance with Lifestyle Medicine

I don't understand why asking people to eat a well-balanced, vegetarian diet is considered drastic, while it is medically conservative to cut people open.

Dr. Dean Ornish
A physician and well known for his lifestyle-driven approach to the control of coronary artery disease (CAD) and other chronic diseases

It is a series of complicated events. It is so easy for a doctor to prescribe a patient a pill to clear up their symptoms; but what about removing the underlying cause of the disease? Surely it should be the doctor's professional job to give the best available health information he possesses, so that patients can decide for themselves which possible, optional, conventional or natural treatments are best or are out there to help their condition.

In this busy modern era, doctor's may not have the inclination, nor time, to educate and clearly explain, to every single-one of their patients – about the importance of changing one's diet and lifestyle, including the long-term health impact lifestyle medicine will have on their health. Of course, if the doctor hasn't learned about the powerful healing effects of natural herbs and plant foods during his medical school days, he obviously cannot possibly pass on such important knowledge in a correct way to any of his patients.

Caldwell B. Esselstyn Jr, Gene Stone, and T. Colin Campbell thus wrote in their book, "Forks over Knives":

> Sadly, today most doctors treat only the symptoms not the causes, giving their patients' pills, carry out procedures and operations that are expensive and dangerous. Some of them say, "Well, patients won't change their diet". However, if you sit with patients for five hours, as we do, and explain exactly what diet does, why, and how, we find that patients do change successfully. This kind of care restores the covenant of trust that must exist between the caregiver and the patient, who realizes you are telling him or her absolutely everything you know about the disease, so it can be either halted or maybe even reversed, so that the patient then feels empowered to do so.[1]

The other issue critics bring up is that people will not stick to a new diet. To this too, Esselstyn, Stone and Campbell have a ready reply:

> When you make big changes in your diet, your health improves quickly. Our clients tell us that they no longer want cheese, cookies, and cakes anymore. Going backward to bad food choices means going back to the past in taking medications and even spending time in hospital. They would rather be healthy.[2]

Synthetic pills and drugs actually fill doctors' pockets with money, and generate billion-dollar profits not only for pharmaceutical companies, but also for the processed foods, meat and dairy industries. We must ask ourselves: "What is more valuable and important: population health or political and financial profits?" In today's world, it seems that for our health care systems, our governments, most of our doctors, the drug and food industries, the absolute primary focus is to make lots of money; human health, unfortunately, comes only secondary.

Considering that most medical research goes to make synthesized drugs that are intended, designed and marketed solely for money, then it comes as no surprise that people think that there is "no money to be made" from selling or patenting natural foods that can benefit human health.

Do People Actually Know About Alternative Cancer Treatments?

The reason why many of us have not yet heard of the alternative life-saving cancer therapies is first and foremost because any natural, chemical remedy or food found in nature that cannot be patented will undoubtedly not sell nor make big billion-dollar profits for the drug and meat/dairy cartels. For this reason, such a cure will be considered impossible and financially not viable.

Besides, many medical practitioners do not believe that a healthy diet can treat any form of chronic disease like medical drugs can. Until new food laws and policies are enacted, with the specific intent to focus primarily on nutrition excellence, and unless it is acknowledged that plant foods can actually prevent, control and many-times reverse diseases, then no health care system will prevail in effectively assisting and aiding patients with chronic diseases.

What Medical Professionals do not "Know or Believe", they may Certainly not Use for Cancer Treatment

There *are* certain health situations when we do require a doctor or physician, and such situations are many. Indeed, doctors should be commended for their contribution in such situations, especially in cases of life-threatening infections, accidents, brain aneurisms, strokes, blood transfusions, obsessive bleeding, burns, cuts, broken bones and other life-threatening incidences.

As former cancer patient Beata Bishop remarks in her book, "A Time to Heal":

> Conventional treatment is truly remarkable for dealing with an emergency and acute cases of disease: it saves lives! It is fantastic! However, when it comes to chronic-degenerative conditions, they do not know how to deal with it.[3]

Medical doctors perform best in emergency situations and which acute problems; this is their greatest value to society.[4] I deeply admire them for this.

So, the medical profession is effective and uses advanced treatments in the case of many acute health care situations. But, when its health care relates to chronic diseases – atherosclerosis, other heart-related diseases, diabetes, auto-immune diseases, obesity, bone diseases and cancer, which is the second leading cause of death worldwide – our health care system is found unresponsive, unworthy and are truly lacking expertise. Why?

Do our doctors and physicians really know or believe that sugar feeds cancer? Do they know that cancer prefers an anaerobic and acidic environment? That animal produce is one of the leading contributing factors for the development of cancer and other degenerative diseases? That cow's milk is detrimental to humans and should be only drank by baby cow's? That specific plant foods; in the long-term can actually achieve better results than conventional therapies; like surgery, chemotherapeutic drugs and radiation as anti-cancer treatments in the inhibition, progression, and regression stages of cancer?

Some of the answers to these questions are unknown to many health professionals, because the latter *does not believe*, or *do not know* about, the healing power of nutrition.

But if we were to talk to doctors about the latest drugs on the market, then it would be a different story. We need to train all health-care professionals so that this may someday change the way our health care system works. Let us treat cancer patients with diet and lifestyle changes and, maybe, combine these with some conventional treatments for the management treatment for all types of non-communicable diseases.

Again, it is important to consider and remember that if doctors, in their medical school days, have not been exposed to the imperative health benefits of diet and clinical nutrition and other alternative (natural) cancer therapies; how then, can they possibly recommend such *Naturopathic conducts* to aid their patients? Perhaps doctors are not entirely at fault here, as they have not been sufficiently trained about such therapies.[5]

Until the medical profession admits, or rather believes, that lifestyle medicine is the primary medicinal remedy for the prevention and treatment of chronic diseases, there can be no future nor hope for bettering chronic medical care. If doctors start acting, start accepting and start believing this, then chronically-ill patients can stand a better chance to change their lifestyle and maybe cure themselves from their own disease.

Notes for chapter 13

1   Esselstyne Jr, C.B., Stone, G., & Campbell, T.C. (June 2011). *Forks Over Knives: Part 1: The Plant-Based World of Forks over Knives.* (1st ed) pp.18-19. Published by The Experiment.

2   Ibid.

3   Bishop, B. (June 1st 1999). *A Time to Heal: Triumph over Cancer, the Therapy of the Future.* (2nd ed). Arkana series, Published by Pengiun UK.

4   Richards, J. B. (2006). *Fight for Your Health: Exposing the FDA's Betrayal of America. Chapter 20: Whatever Happened to Thining Doctors? The Take Charge series*, p. 183. Published by Truth in Wellness, LLC.

5   Adams, K.M., *et al.* (April 2006). Status on nutrition in medical schools. *The American Journal of Clinical Nutrition*, **83**(4), pp.941S-944S. Parker, W.A., *et al.* (August 11th 2011). They think they know but do they? Misalignment of perceptions of lifestyle modification knowledge among health professionals: *Public Health Nutrition*, **14**(8), pp.1429-1438.

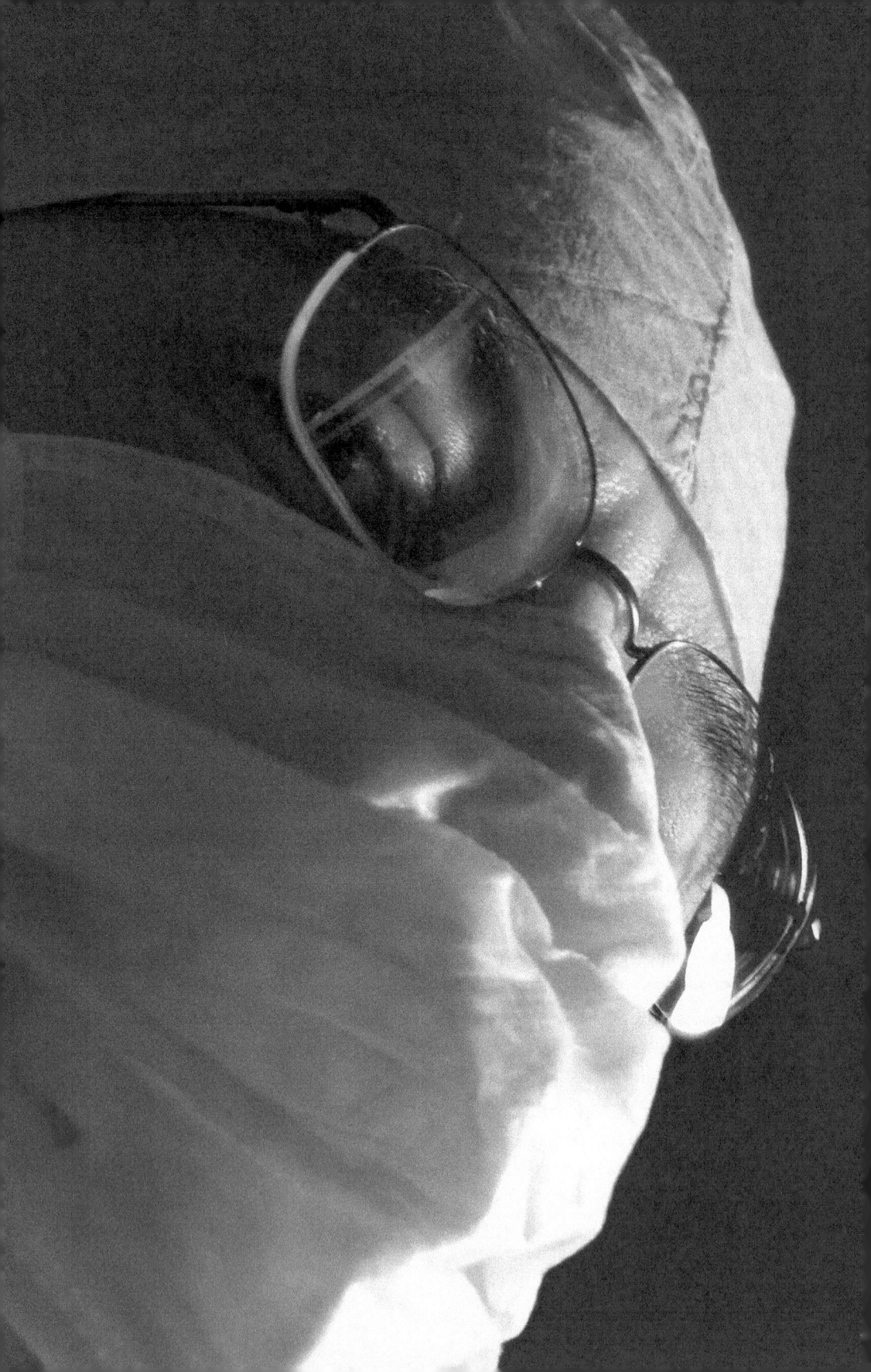

14

Medical Malpractice

The patient does not care about the science; what he wants to know is, can you cure him?

Martin H. Fischer (1989-1962)
German-born American Physician and Author

Look to your health – and value it next to a good conscience; for health is... a blessing that money cannot buy.

Izaak Walton (1594-1963)
English Writer

Most health professionals still do not adequately emphasize, and most unfortunately, *still* give patients the "permission" to continue eating the same diets that had caused them their problems. So, the unacceptable consequence is that the underlying disease is being ignored.

Governments and health care systems should spend much more time and money towards empowerment, awareness and in educating patients about nutrition excellence. Generally, medical professionals do not insist enough with ill patients about the correct dietary solutions. Improving doctors' and patients' compliance with a "Lifestyle Intervention Plan" will most certainly bring about a far superior way in dealing with patients' self-care problems and improving or reversing diseases. Instead, we are seeing patients who have already been treated for heart attacks, strokes or cancers, returning for medical care or hospitalization, once again experiencing the same symptoms they may have had six months before. Now, this may be considered as "malpractice."

Malpractice is, after all, the failure of a professional person, as a physician or lawyer, to render proper services through reprehensible ignorance or negligence or through criminal intent, especially when injury or loss follows.[1]

All Drugs Have Side-Effects: Mental and Physical

What most doctors do when treating today's diseases is doing little to improve human lives. Indeed, in many cases, their contribution is ineffective. Why is this so? Because the drugs prescribed by doctors encourage patients to continue with their risky lifestyles and self-destructive eating habits. They give patients "permission" to continue with their errant behavior because they mask the symptoms of the disease, as already stated earlier in this book.

So, when considering the risks of any drug, medical treatment or intervention, one needs to consider the benefits of lifestyle interventions, such as exercise, salt avoidance, dietary modifications, and weight reduction. Such interventions have no side-effects and focus on removing the cause; they do not just treat the symptoms of the disease.[2]

If the underlying causes are addressed, patients are often able to stop taking medication (under the doctor's supervision, of course). In certain cases, they can even avoid surgery. Medical care spends billions cracking people's chests open or cutting out tumours, but only rarely does it actually prolong anybody's life. How about trying to wipe out at least 90 percent of heart disease? Just think about it. Heart disease accounts for more premature deaths than any other illness, and is almost completely preventable by simply changing one's diet and lifestyle. The same changes can prevent or reverse many other chronic diseases.

So why don't doctors finally acknowledge it and start using lifestyle medicine in their cancer treatment protocol?[3]

One reason could be that doctors do not get paid for it. "No one profits from lifestyle medicine," one could say. Lifestyle medicine is not part of medical education or practice. Presently, however, physicians lack the training and the financial incentives that could encourage them to teach patients about the benefits of a healthy diet, exercise, stop smoking, managing their weight and addressing the importance of human detoxification. So physicians continue to do the only thing they know: prescribe medication and perform surgery.[4]

An enormous body of evidence supports the effectiveness of lifestyle interventions for lowering the risk of developing chronic disease, as well as for assisting in the management of existing disease.[5] When comparing conventional and lifestyle interventions, lifestyle medicine truly has the edge and "long-term advantage", when dealing with chronic diseases. It also lets people take control of their own health. If a good health care team work together and patient compliance is made, lifestyle medicine can move forward, and show its true brilliance in increasing patient health and long-term survival.[6]

Dr. Dean Ornish proved that we could reverse the No. 1 cause of death, that is, heart disease. He could open up arteries not with drugs nor surgery, but just with a plant-based diet and healthy lifestyle changes. He thought his studies would have a "meaningful effect" on mainstream cardiology.[8] He admits, however, that he was mistaken. Why?

Physician reimbursement, he realized, is a much more powerful, determining factor of medical practice than research: reimbursement comes before research, salary before science, wealth before health. Indeed, not a very flattering portrayal of the healing profession, unfortunately. So, if doctors won't do it without getting paid, why not get them paid?

Dr. Ornish went to Washington arguing: if we train and pay doctors to learn how to help patients address the real cause of disease with lifestyle medicine, and not only treat diseases, we could save millions of patients' lives – and that is just talking about heart disease, diabetes, prostate and breast cancer!

The Take Back Your Health Act was before the U.S. Senate to encourage doctors to learn and practice lifestyle medicine: both because it works better for the patients and also because physicians will get paid for it. Sadly, the Bill met its death... just like the millions of Americans will continue to do from deadly chronic diseases.[9]

Until about half a century ago the most common causes of death throughout the world were infectious diseases. These included not only the great epidemics of tropical diseases, but also infections such as whooping cough, tuberculosis, scarlet fever, measles and gastroenteritis. In the economically more developed communities, this is no longer the case. It can no longer be disputed that a formidable list of diseases, common in affluent societies is, in fact, characteristic of modern Western culture. Moreover, an impressive volume of evidence has now been accumulated indicating that diets are the major, though not the exclusive, cause.[10]

Life Expectancy

Exposure to medical care and the resources spent on health care are actually linked to decreases in healthy life expectancy, not increases. Emergency hospitalization care is valuable, but in the modern world emergencies linked to injury, accidents, and infections are no longer the leading causes of death: Heart diseases, strokes, and cancer have become "the big three".

The same goes with respect to all other health care practitioners and health care establishments who may have been deceived, distracted from, or blinded to the truth. Most medical practitioners and health care systems unfortunately know of only one way to treat all forms of chronic diseases: the "conventional way".

When we talk about natural plant foods, we are talking about eating broccoli, kale, flax seeds, Brussels sprouts, spinach, cauliflower, beetroots, carrots, red cabbage, tomatoes, onions, garlic, leeks, many types of mushrooms, fruits especially berries, nuts, seeds, legumes and whole grains. Indeed, there is not a lot of money to be made from the many hours of scientific research by the cancer industry on the use of these specific anti-cancerous foods in the treatment of cancer patients. Yet, cruciferous vegetables like kale, broccoli, Brussels sprouts and boy chow and other plant-derived foods have many incredible medicinal healing-powers.

A doctor can give you a pill that lowers or raises your blood pressure; a pill to help your heart beat slower or faster; make you urinate more or urinate less; or to lower your blood sugar levels. Then again, we can go to a health-food store and just buy healthy, natural foods or herbs that can be just as effective!

However, the efficiency of pharmacological substances, whether natural or not, is proportionate to their toxicity. They do not work or do they actually repair anything; they work because they actually try to balance, block or interfere with the natural body defenses. In most cases, the symptoms of disease are the body's natural ability to try to repair or rejuvenate its defenses against the disease. Plant foods can do this naturally, without leaving any long-term side-effects in the body. We need to change how doctors, physicians, our health-care systems and medical foundations work.

Lifestyle medicine is the best alternative and natural way that can improve the health of chronically-ill patients. Patients will not be left stranded with unnecessary surgical procedures; they will not have to take drugs or medications for the rest of their lives; they will not have to suffer excruciating pain nor inflict unnecessary anguish and sorrow on their families lives. These immoral and unethical circumstances may all be prevented and treated with lifestyle medicine.

Notes for chapter 14

1   Dictionary.com. Malpractice. Retreved from http://dictionary.reference.com/browse/malpractice?s=t

2   Hayman, M.A., *et al.* (2009). Lifestyle Medicine: Treating the Cause of Disease. *Alternative Therapies,* **15**(6), pp.12-14.

3   Greger, M. (November 4th 2013). *Lifestyle Medicine. Treating the cause of disease.* Volume 15, video. Retreived from http://nutritionfacts.org/video/lifestyle-medicine-treating-the-causes-of-disease/

4   Ibid.

5   Rippe JM, Crossley S, Ringer R. (1998). Obesity as a chronic disease: modern medical and lifestyle management. *J Am Diet Assoc,* **98**, pp. S9-S15. Nonas CA. (1998). A model for chronic care of obesity through dietary treatment. *JADA,* **98**, pp. S16-S22. Rippe, JM, Angelopoulos TJ, Zukley L. (2007). Lifestyle Medicine Strategies for Risk Factor Reduction, Prevention, and Treatment of Coronary Heart Disease: Part II. *American Journal of Lifestyle Medicine,* **1**(2), pp.79-90. Svetkey LP, Erlinger TP, Vollmer WM *et al.* (2005). Effect of lifestyle modifications on blood pressure by race, sex, hypertension status, and age. *J Hum Hypertens,* **19**(1), pp.21-31. Eddy DM, Schlessinger L, Kahn R. (2005). Clinical outcomes and cost-effectiveness of strategies for managing people at high risk for diabetes. *Ann Intern Med,* **143**(4), pp.251-64. Rimm EB, Williams P, Foster K, *et al.* (1999). Moderate alcohol intake and lower risk of coronary heart disease: meta-analysis of effects on lipids and haemostatic factors. *BMJ,* **319**, pp.1523-1528. Sato Y, Nagasaki M, Kubota M, Uno T, Nakai N. (2007). Clinical aspects of physical exercise for diabetes/metabolic syndrome. *Diabetes Res Clin Pract,* **77**(Suppl 1), pp.S87-91. Giannattasio C, Mangoni AA, Stella ML, et al. (1994). Acute effects of smoking on radial artery compliances in humans. *J Hypertens,* **12**, pp.691-691.

6   Lifestyle Medicine –Evidence Review. (June 30, 2009). American College of Preventive Medicine, p.1-70.

7   From Eggar, *et al.* (2004). Lifetsytle Medicine. Sydney: McGraw-Hill, p.4

8   Ornish, D., *et al.* (1990). Can lifestyle changes reverse coronary heart disease? The Lifestyle Heart Trial. *The Lancet,* **336**, pp.129-133.

9   U.S. Congress S.1640, 111th. (August 6th 2009). *A Bill: To amend title XVIII of the Social Security Act to provide coverage of intensive lifestyle treatment.* Take Back Your Health Act.

10 Burkitt, P.D. (1982). Lifestyle and Disease; Western diseases and their emergence related to diet. S. *Alr. Medical Journal,* **6**, pp.1013-1015.

15

Guidelines on How to Deal with Cancer

Time and time again it has been confirmed that the proven medical treatments are not only ineffective but dangerous. The vast majority of patients with cancer live longer and better if left without the orthodox treatments.

Francisco Contreras, M.D.
Director, president and chairman of the Oasis of Hope Hospital. A distinguished oncologist and surgeon.

Do Not Fear Cancer, But Fight Against it with Diet and Lifestyle

Because you are reading this, you no doubt have cancer, or know someone who does. With such a serious disease, not everyone gets well, no matter what program they follow. But if you follow the rules of a strict diet and lifestyle, your chances of healing cancer will skyrocket, compared to if you do not follow these guidelines.

In contrast, our current medical model uses powerful poisons to rearrange the body's chemistry in order to suppress *symptoms* of the disease. It also cuts out or surgically alters organs that are causing problems, instead of healing the organs. The amazing ability about diet and lifestyle changes is that you don't have to know a thing about cancer to reverse it. You just have to know a thing or two about what you put into your mouth and how to live a better life.[1] Another purpose of this book is to tell you a thing or two about how to accomplish both, without using toxic treatments that impair the body and senselessly or negatively thinking there is no hope of survival without today's so-called conventional treatment. You can become your own doctor and try to heal cancer yourself.

But, if you already have breast cancer, fear and anxiety can undoubtedly cloud your judgment. This could lead you down a road. A road that always looks dark, lonely and empty, with no bright lights ahead. But there is hope. A hope for a change and that change can open up that road and bring in a full spectrum of opportunities, where you will feel more confident with yourself in knowing that there may certainly be such a cure on that glooming road of hopefulness, to a newer and brighter path to recovery.

Even though the cancer may have been with you for decades, a microscopic tumour that could not be felt or recognized, may suddenly develop and become large enough to be medically diagnosed.

Your doctor could just tell you that you must undergo conventional medical treatment immediately in the form of surgery, chemo and/or radiation. He might even use scare-mongering tactics to try to force you into treatment. "You may die if you don't start treatment immediately" are the usual words.[2]

The following vital, initial steps are what Mike Anderson in his best-selling book "Healing Cancer from inside out" and I would strongly suggest and consider being your best option:

1. Continue to read the rest of this book and others similar to it.
2. Start a plant-based diet and improve your lifestyle straight away. Please do not wait. It does not hurt you; it will only help. The health implications are all beneficial and the treatment is superior, and leaves no side-effects.
3. Choose all the anti-cancer foods that other physicians and experienced dieticians understand and that I have suggested in this book. The many years of hard work and research has already been done for you. The studies have all been documented and have shown that *specific* plant foods can attenuate the initiation, promotion and the progression stages of the disease.
4. Always get a second opinion, a second diagnosis. For example, breast cancer diagnosis can be problematic: read chapter 20 about mammogram screening.
5. Get your markers re-checked after a few months. If you are making progress and the cancer has reduced in size (showing that you are winning and outsmarting your battle with cancer) then – Please! Please! Please! – Be strong and determined, and keep on the anti-cancer diet and stay the course.
6. If you are not making progress, then you have more serious decisions to make. You may be in need of specialized treatments, in which case I would advise you to seek professional help from a clinic which has experience in reversing cancers using natural treatments.[3] Alternative clinical cancer treatments and resources for these treatments are given at the end of this book.

Alternative treatments *do not* harm healthy human cells, and focus more on eradicating malignant tumour cells.

This is a key difference between alternative cancer treatment protocols, and conventional cancer therapies. The latter are not toxic-selective, that is, they kill *all* cells, including healthy cells.

Alternative cancer treatments focus more on cleansing the body and stimulating the body's natural defenses (including the human immune system) with plant-based diets, supplementation, oxygenation and detoxification and other lifestyle interventions.

There are numerous alternative cancer therapies worldwide that rely on one supplement or another or on one specialized type of treatment. However, instead of focusing on supplementations, this book focuses on nutrition because many experienced medical professionals, whose opinion I share "believe that nutritional excellence is the foundation for reversing cancer", regardless of supplementations. There are specific minerals and vitamins that, when taken in small or high doses (for example, the mineral selenium, Vitamin C, D and Vitamin B_{17} or Laetrile), can be most helpful for the treatment of cancer.[4]

Go all the way!

Supplements and specialized cancer treatments cannot reverse cancers by themselves, particularly in the long-term. Basically, if you are serious about reversing cancer, you should be strongly and very determined about changing your diet and lifestyle. There can be no moderation in the diet. You must go 100 percent! You must go all the way when you are suffering from a long-term chronic disease. Diet and lifestyle changes are the real foundation of reducing your risk of all forms of diseases: diet and healthy lifestyle changes will certainly start the healing process and keep cancer at bay.

Can Dietary Treatment Work When Undergoing Conventional Treatments?

In most cases, natural healing therapies can be far superior for cancer treatment if one has had no previous orthodox therapy. Why is this?

Over 95 percent of all cancer patients who seek nutritional treatments have already had extensive conventional cancer treatments. Following such conventional treatments, many of them would have already been sent home to die.

Consider the typical condition of these patients before they sought alternative treatments:[5]

1. The patient's immune system has already been virtually destroyed by chemotherapy.
2. Poor immunity weakens the body, leaving it vulnerable to attacks from microbes. This makes the body highly acidic and oxygen-depleted, creating the perfect environment for cancer occurrence.
3. The cancer has become stronger than before, as cancer cells have developed a resistance to chemotherapy.
4. Chemotherapy has permanently damaged or impaired, even non-cancerous cells and organs.
5. At least one of the major organs (usually the liver) has been severely damaged.

6. The digestive track is damaged and cannot absorb many of the nutrients from foods.
7. Cancer patients are in extreme pain and, in many cases, have lost their will to survive.

On top of all this, the window of opportunity for nutritional treatments has been largely lost – precious time that could have helped the patients rebuild their immune system to fight the cancer would be lost. The irony of all this is that, "patients usually turn to nutrition and alternative treatments as a last resort, after conventional treatments have failed, that is, if the patient dies, the alternative treatment will be blamed".[6]

People often ask, "But *will* diet really help my type of cancer?" A good dietary cancer plan will, for the most part, certainly help all types of cancers. Whether it is a carcinoma, lung, breast, ovarian, endometrial, sarcoma, melanoma, lymphoma and leukemia, etc. "Diet is vital for helping to rebuild the Human-body's defenses". Whenever your immune system is in top-notch condition and not poisoned by conventional treatments, it will know how to deal with every type of cancer. You should not be fearful of the type of cancer, so much as frightened your immune system is not healthy enough to handle it.[7]

The Main Purpose of Dietary Treatments

The purpose of dietary treatments is to rebuild the body's defenses, particularly the immune system, remove harmful toxins, and rebuild the damage done by conventional treatments and by the patient's cancer cells. Only then can the body make its remarkable fight-back to eradicate all, but now visible, cancer cells.[8]

With respect to cancer, there are four "laws" that are irrefutable:

1. A weakened immune system is the biggest single cause of cancer.
2. Your immune system is the only cure for cancer.
3. The higher the dietary fat, the more the immune system is weakened.

4. The only foods that contain significant immune-strengthening and cancer-fighting nutrients are "Plant foods".

Let's be More Specific and Only Target Cancer Cells

Conventional treatments do not target cancer cells only, but they also target healthy human cells and organ systems, including the most fundamental protective system in our body: the "Human Immune System".

This is like trying to get rid of a fully-grown elephant in an upstairs room of your house. That is, the elephant is at first hesitant to move, but then it finally decides to leave from the front door. Alas, by this time the elephant will have already caused a lot of damage and practically demolished your whole property. Conventional treatments are not specific, they do not just target the cancer cells; they also destroy the very foundation of our defensive system. The immune system would have been virtually wiped-out by chemotherapy, and this essential

protective system, would not be able to function adequately, in order to safeguard us from cancer nor from the recurrence of cancer.

Ty Bollinger succinctly puts it this way: "Any treatment that does not address the underlying causes for the breakdown of the immune system will be palliative at best, and life-threatening at its worst."[10]

To sum up, dietary treatments can help even during conventional treatments, but when applied after such treatments they become a hard up-hill battle. Conventional medicine renders healing-by-diet much more difficult. Without a doubt, your chances of long-term survival will be improved if you do not first opt for conventional treatment, but focus on following a complete anti-cancer plant-based diet. Eating a healthy, specific plant-based foods may actually prevent disease, reverse, and in most cases, may even cure disease, especially for those patients diagnosed with cancer.[11]

Notes for chapter 15

1. Anderson, M. (2009): Healing Cancer from Inside Out. pp.5-6. www.RareDiet.com.
2. Ibid.
3. Ibid.
4. Ibid.
5. Anderson, M., ibid., pp .50-52.
6. Ibid.
7. Anderson, B. (2007) *Cancer-Free. Your Guide to Gentle, Non-Toxic Healing. Clients' Feedback pages*. Published by Booklocker.com, Inc.
8. Ibid.
9. Ibid.
10. Bollinger, Ty. The Cancer Truth. *Cancer does NOT have to be a death sentence. What is Cancer?* Retreived from http://www.cancertruth.net/test-2/
11. Donaldson, S. M. (October 20, 2004). Nutrition and Cancer: A review of the evidence for an anti-cancer diet. *Nutrition Journal*, **3**(19), pp.1-21. doi:10.1186/1475-2891-3-19; Peetha, A, et al. (2008). Cancer is a Preventable Disease that Requires Major Lifestyle Changes. *Pharmaceutical Research*, **25**(9), pp. 2097-2112. DOI: 10.1007/s11095-008-9661-9; Hyman, A.M. (2009). Lifestyle Medicine: Treating the Cause of Disease. *Alternative Therapies*, **15**(6), pp.12-14; Vay Liang, et al. (2001). Diet, Nutrition and Cancer: Where Are We Going from Here? American Society of Nutritional Sciences. *Journal of Nutrition*, **131**, pp. 3121S-3126S; Val Liang, W. Go, et al. (2003). Diet, Nutrition, and Cancer Prevention: The Postgenomic Era. American Society for Nutritional Sciences. *J. Nutr.* **133**, pp.3830S–3836S; Ricardo Uauy and Noel Solomons. (2005). Diet, Nutrition, and the Life-Course Approach to Cancer Prevention. *J. Nutr*, **135**, pp.2934S-2945S.

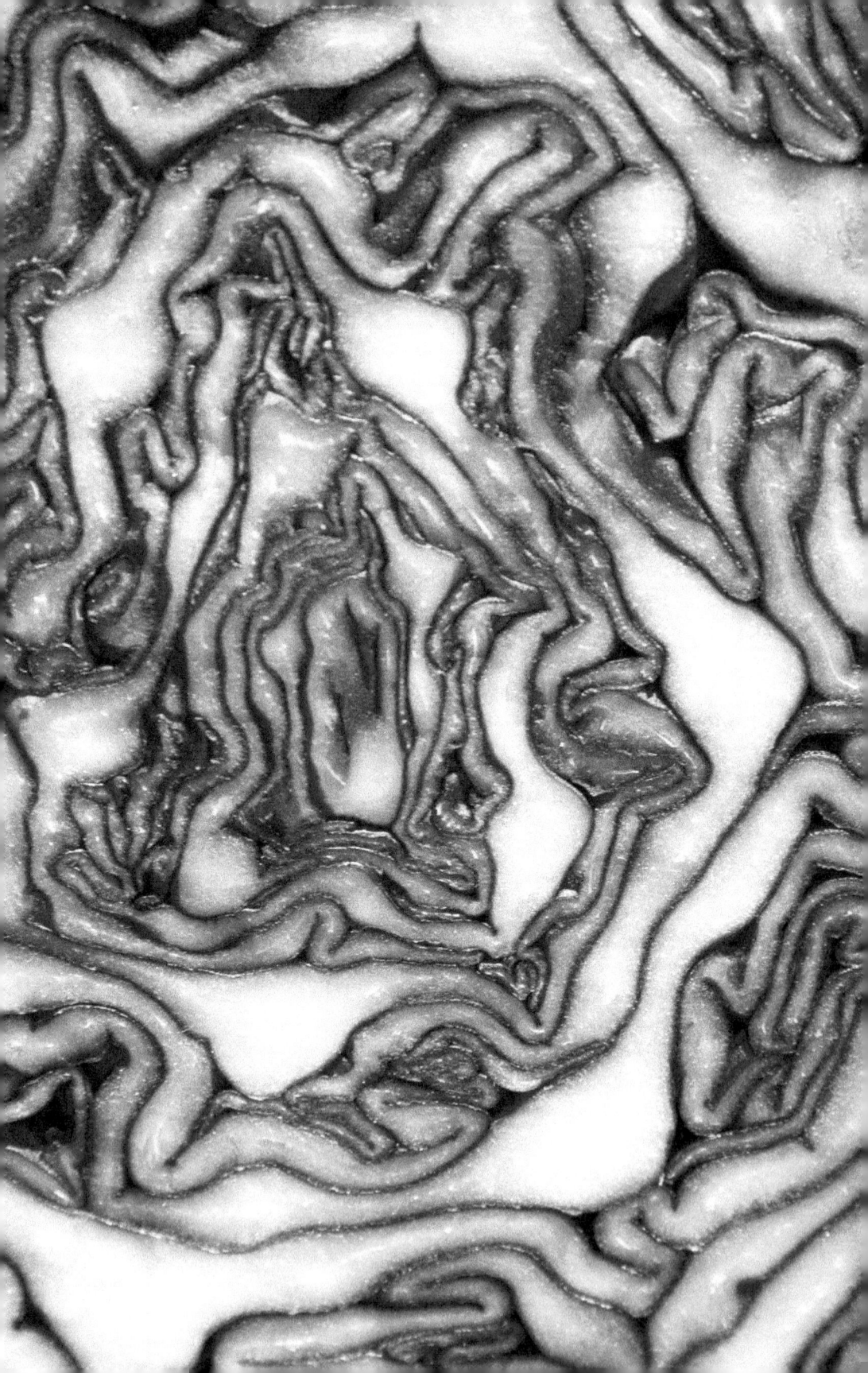

16

The Health Benefits of Phytochemicals

Plants are nature's alchemists, expert at transforming water, soil and sunlight into an array of precious substances, many of them beyond the ability of human beings to conceive, much less manufacture.

Michael Pollan
The Botany of Desire: A Plant's Eye-View of the World

Phytochemicals *are* Powerful Healing Compounds

Every day, scientific reviews and journals document the healing powers of phytochemicals (Phyto means "plant") which began to be discovered by scientists in the past decades. These plant components are essential and extremely beneficial to human health. Phytochemcials are miraculous healing compounds that can significantly help towards the prevention and reversal of disease.

The most common and most natural feature of plants is also their most obvious features, which is their wide range of bright-pigmented colors. If you admire how food is presented, it is hard to beat a plate full of fruits and vegetables: the greens, blues, reds, purples, yellow and orange colors of plant foods are indeed very tempting.

The link between beautifully colored plants and their exceptional health benefits has often been noted. There is a beautiful, scientifically sound story behind this color/health link. The colors of plant foods are derived from a variety of chemicals called "antioxidants". These chemicals are almost exclusively found in the plant kingdom, although some are also found in animal-based foods – because whenever animals eat plants they store a small amount of these antioxidants in their own tissues. The consequences of not consuming sufficient amounts of such chemicals are even more astonishing: premature death, cancer, and a number of heart and other diseases. Chronic diseases may develop if these chemicals are not obtained from plant foods. But why, and what, is so special about plants vis-a-vis human health?

Antioxidants are a class of phytochemicals, vitamins and minerals which have been found to play protective roles not covered by carbohydrates, fats and proteins, that include the following:[1]

1. Controlling the production of free-radicals
2. Protecting cell structures from damage to toxins
3. Inducing beneficial antibacterial, antifungal and antiviral effects
4. Impeding the replication of cells with DNA damage
5. Inhibiting the function of damaged or genetically altered DNA

6. Fuelling mechanisms to repair damaged DNA sequences
7. Inducing detoxification enzymes and helping the liver's 2-phase detoxifying system
8. Boosting the human immune system's (destructive cytotoxic T-Killer cells) power: that is, the protection and power to kill microbes and cancer cells
9. Deactivating and detoxifying cancer-causing agents.

The above list could be condensed into one primary role,

"Phytochemicals, many of which also work as antioxidants, are the fuel that assists the human body's anti-cancer defenses."

But, there is a problem. Humans *do* have a detoxification process: the human liver's Phase I and Phase II detoxification system. To simplify things:

Phase I either directly neutralizes a toxin, or modifies it (partially biotransforms it); thereby creating an intermediate or metabolite of the original chemical compound, which are then neutralized by one of the several Phase II enzyme systems. In Phase I, a toxic chemical is converted into a less harmful chemical: this is highly achieved by various chemical reactions (such as hydrolysis, reduction and oxidation), and during these reactions free radicals are produced which, if these accumulate, may damage human liver cells. The primary function of phase II is to further biotransform compounds to a less toxic, more hydrophilic compound. This is where the importance of plant foods and their antioxidant potential play their vital role. If essential plant food compounds are lacking and the toxic exposure is high, toxic chemicals become highly dangerous. Some may be converted from relatively harmless substances into potentially carcinogenic substances. Many of these are chemicals are then metabolized in the human liver, and their byproducts are then eliminated from human cells and find their way into either the urine or the bile.[2]

But, human beings do not have a naturally-built shield, which protects us against harmful chemicals and free-radicals. We are not plants, and we do not carry out photosynthesis, and therefore, the human liver can only produce small amounts of these essential and protective compounds, including antioxidants. However plants have the amazing ability to produce thousands of them. So it is said that we need to aid and stimulate the function of the human liver, with the consumption of phytonutrients: that may ultimately help boost the liver's Phase I and Phase II detoxification system, in order to convert any excess harmful toxins into less harmful ones.

Colleen Patrick-Goudreau, in "Color Me Vegan", writes:

> The phytochemicals, antioxidants, and fibre – all of the healthful components of plant foods – originate in plants, not animals. If they are present in animals, it is because the animals have eaten plants. And why should we go through an animal to get the benefits of the plants themselves? To consume unnecessary, unseemly, and unhealthy substances, such as saturated fat, animal protein, lactose, and dietary cholesterol, is to negate the benefits of the fibre, phytonutrients, vitamins, minerals, and antioxidants that are prevalent and inherent in plants.[3]

Very fortunately for us, the antioxidants in plants work in our bodies the same way they work to protect plants from free radical damage. It is a wonderful harmony. Plants make the antioxidant shields, and at the same time make them look incredibly appealing with amazing, beautiful and appetizing colors. Then we, animals, in turn, are attracted to the plants, and whenever we choose to consume these healthy, edible, non-toxic plant foods, we then also borrow their antioxidant shields for our own health. Whether we believe in a superior being (God), evolution, or just coincidence, we have to admit that this is a harmonized, beautiful and almost spiritual example of nature's wisdom.[4]

A Plant-based Diet Rich in Phytochemicals is the Best Way to "Prevent Cancer."

Whenever we consume a variety of vibrant plant foods, this creates a defensive shield around our cells and tissues, and this shield can be considered to be colored mostly green, yellow, blue, purple, red, orange, and many other vibrant colored plant foods. Without the array of these wonderful colors there would be no beneficial power within plants. These abundant amounts of phytochemicals will ultimately shield us from the development of disease.[5]

Within the next 10-20 years this will no longer seem a myth to anyone. Everyone will or should know of the important and protective power of these chemicals. Phytochemicals are keeping us alive by protecting us from the harmful rays of the Sun, the exposure to harmful toxins and other environmental carcinogens.

We humans have collectively polluted our environment with toxins. Plants are the only real source of protection against these toxins. We should eat plant vegetation as if our lives depended on them.

I believe that the medical practice in the twenty-first century will focus less on the symptoms of disease – considering the internal, metabolic, physiological and mental damage that drugs, that are foreign to our bodies, can cause us. A healthy (plant-based) diet focuses more on empowering the body from within, boosting it by the disease-fighting powers of the antioxidant network found in phytonutrients.

This is why plant foods are so essential. Humans certainly have to eat a variety of plant foods if they want to get all the required phytochemicals that are needed to enable all body cells to work in synergy with one another, and to offer overall protection against all known degenerative diseases.

Dr. Dean Ornish's Advice

Dr. Dean Ornish published an editorial in a 2010 edition of the American Journal of Cardiology that really sums up the stage that has been reached today.

He describes a growing body of scientific evidence that the best diet is a plant-based diet, consisting predominantly of fruits, vegetables, whole grains, legumes, soya and lentil products. He also suggests the inclusion of nuts and plant roughage (fibre). If we do as he suggests, we can do away with our bad cholesterol, do away with all the drugs we consume our whole life – and all this without the costs and the side-effects attached to drugs.[6]

Of course, most patients are not even given this healthier option; because of the erroneous belief among doctors that patients will not stick to such diet anyway, or that a plant-based diet will not bring about any benefit to human health.

But, in reality, most people do not want to take the drugs. When patients are asked to take pills, even with the intention of preventing or curing serious diseases, most people's first reaction is to avoid taking such pills.

However, Dr. Ornish notes that:

> When people make comprehensive lifestyle changes, including a plant-based diet (or a modified plant-based diet), they often feel so much better, so quickly, that it reframes the reason for making these changes from fear of dying, which usually is not sustainable, to joy of living, which often is.[7]

A good diet, unlike drugs, does not just help *lower* cholesterol levels. A good diet can also *prevent* a variety of chronic diseases, including: hypercholesterolemia (high cholesterol levels in the blood), coronary heart disease, diabetes, hypertension, obesity, prostate cancer, and breast cancer.

Pills cannot do all that. In the world of drugs, you cannot find a drug specifically for heart disease, and another drug specifically for diabetes. But a plant-based diet can cover the entire spectrum, i.e. all medical conditions. Plant-based foods contain more than 100,000 disease-preventing nutrients.

It is important that we say that again. *Yes, there are 100,000 different disease-preventing nutrients that can be found in plants.*[8]

Just to name a few, blueberries contain phytochemicals called anthocyanins that can improve one's memory. Tomatoes are rich in the pigment lycopene, an antioxidant that can reduce the risk of coronary heart disease and prostate cancer. Ginger contains a compound called gingerol that lowers blood pressure and enhances blood circulation. Pomegranates contain cyanidin, pelargonidin, punicalagin, punicalin and ellagitannins that may have anti-proliferative cancer fighting nutrients. However, let us not forget the phytonutrient power of vegetables like the life-saving and health-boosting broccoli, kale, spinach, red cabbage, cauliflower, garlic, leeks, beets, onions and all the other members of cruciferous and the Allium family of vegetables. These types of vegetables have been scientifically proven to inhibit and reduce cancer growth. Indeed the list of photochemical and nutritional benefits that can be obtained from a plant-based diet goes on, and on, and on...

However, when these phytochemicals were taken in pill form, beta-carotene supplementations were found to increase the risk of lung cancer in smokers.[9] On the contrary, foods such as whole carrots, that are naturally rich in beta-carotene, were found to lower the risk.[10] Anyway, we cannot possibly swallow 100,000 pills a day to resolve all our acute or chronic medical conditions!

The "take-home message": if you want vitamin E, vitamin C or beta-carotene, don't reach for the pill bottle – reach for the leafy green vegetables or fruits. Ornish also talks about his efforts in slowing, stopping and reversing the progression of severe coronary artery diseases and even cancer. Living and eating healthy actually changes us on a genetic level – up-regulates disease-preventing genes, and down-regulates genes that promote breast cancer, prostate cancer, inflammation and oxidative stress.

Drugs Cannot do What Plant-Based Diets Can

Dr. D. Ornish concludes by stating that:

> Many people tend to think of "breakthroughs in medicine" as new drugs, lasers or high-tech surgical procedures. They often have a hard time believing that the simple choices that we make in our lifestyles – what we eat, how we respond to stress, whether or not we smoke cigarettes, how much exercise we get, the quality of our relationships,

and social support – can be as powerful as drugs and surgery, and they often are, and sometimes actually are, even better.[11]

This is why no single or multiple supplementation will ever succeed in healing the body of disease unless it has been proven to work when consumed at high doses (Vitamin C). It is important to remember that drugs need to be combined with specific plant foods in order to counteract specific diseases.

Let me say it again: These plant nutrients help all cells and tissues of the human body by reducing oxidative stress; by keeping bad inherited and acquired genes intact; by reducing cell membrane/DNA damage and the aging process; and by removing harmful carcinogens. These physiological and metabolic disturbances can interfere with our body's normal functions and alter DNA expression which can, eventually, lead to gene mutations and the formation of cancers.

I will now mention the best anti-cancer foods that sustain precious phytochemicals that have been studied in medical literature and have been proven to fight the battle against breast cancer. Think of the abundance of antioxidants found in plants as life-saving nutrients. These are just a few of the big important advantages of eating plant foods over eating animal foods.

Green-leafy vegetables are the powerhouses of antioxidants, but there are other plant foods that contain abundant amounts of different antioxidants: fruits, nuts, seeds, whole grains and legumes. Vitamin C, Vitamin E, and selenium are powerful antioxidants that are found abundantly in plant vegetation.

The most comprehensive study on this topic was made on the total antioxidant content of more than 3,100 foods – beverages, spices, herbs and supplements used worldwide. This major study can help us take decisions in hundreds of real-life grocery-store situations that we make all the time; although sometimes it is easy to get lost in the details. Let us take a step back and simplify things a little. What, exactly, does this study say about what we should be eating in general? The first thing they did was to split everything into plant foods and animal foods, as the table below shows:

Antioxidant Content: Plants vs. Animal-Based Foods

	N	MEAN	MEDIUM	MAX
Plant-Based Foods	1,943	1, 157	88	289,711
Animal-Based Foods	211	18	10	100

As is evident, on average, plant foods have 64 times more antioxidants than meat, fish, eggs and dairy products (animal-based foods).

This alone represents a powerful argument in favor of a healthy plant-based diet. Animal foods' maximum is 100. Plant foods go up to 289,000.

The study concluded as follows:

> Foods vary several-thousand fold when antioxidant rich foods originate from the plant kingdom, while meat, fish and other foods from the animal kingdom are low in antioxidants (...) Diets comprised mainly of animal-based foods are, thus, low in antioxidant content, while diets based mainly on a variety of plant-based foods are antioxidant-rich, due to the thousands of bioactive antioxidant phytochemicals found in plants, which are conserved in many foods and beverages.

In another review, on vegetarian diets and public health, the researchers concluded that there is sufficient scientific evidence that supports public health policy that promotes a plant-rich diet for healthy lifestyles. It does not need to wait for science to provide all the answers as to why and how.[12]

But the study, done on 3,100 foods and their antioxidant content, is certainly one strong reason. On average, as we said above, there are 64 times more antioxidant power in plant foods than in animal foods.[13] But is this really a fair comparison?

Could it be that some foods with high anti-oxidant contents were purposely included in the study and compared to animal foods, thus producing a bias in favor of plant-food? Did the plant-food under study include exotic wild berries, a product that you could never find in your local grocery store, thus skewing the chart upwards for the plant-foods? Most people do, indeed, eat foods like corn or potatoes, but they do not normally eat things like dried Norwegian cornflowers or Mexican gooseberries! So, let us be fair and bring the results of the said study down to earth.

The average (mean) plant food does indeed have over 1,000 units of antioxidant power, but for comparative purposes, and to be fair to animal kingdom, let us take the *least healthy* plant food, the good old iceberg lettuce which is, basically just water. The iceberg lettuce certainly does not contain more than 1,000 units of antioxidant power; it only has 17 units.

Although iceberg lettuce contains only 17 units of antioxidant power, it still beats fish (11), salmon (7), and chicken or pork (6). Iceberg lettuce has nearly 3 times more antioxidant power than chicken. A hard-boiled egg has just 2, while egg white has zero. Even Coca-Cola has 4 units, i.e. the same amount found in cows' milk or yogurt – though soy milk only has about twice that amount. While plant foods average over a 1,000, the best animal foods can do (in the meat category of animal-foods) is a serving of ox liver at 71.

Ox liver beats out other animal meats, like moose meat and reindeer steak, but still cannot quite reach the antioxidant power of a "Snickers" bar!

This is why you need to eat a plant-based diet, because even if you lived off ox liver – the "wild blueberry of the animal kingdom", that is, one of the few animal foods that can beat out the iceberg lettuce – you would still never come close to your daily antioxidant needs.

There is one animal product, however, that *does* kick some serious tosh, topping 200. There are even some types of berries that did not result to be that antioxidant-rich. This animal product is so healthy that I have to encourage everyone to consume it – when you are still a baby. And that is the antioxidant content of human breast milk! During

THE HEALTH BENEFITS OF PHYTOCHEMICALS

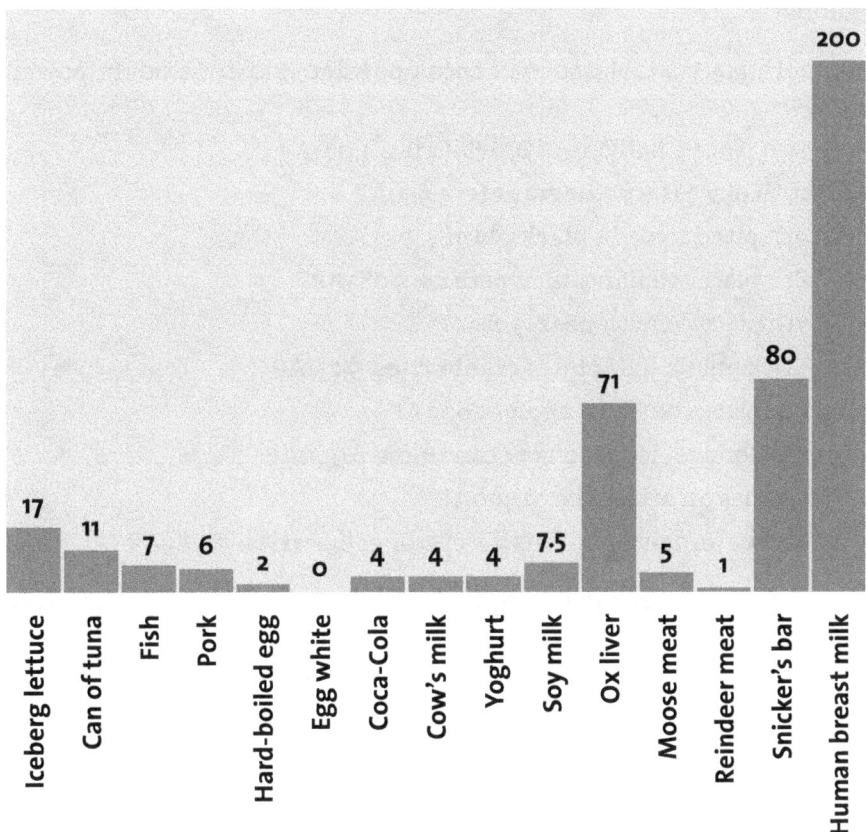

Units of antioxidant power

infancy, breast is best. After infancy, plants are healthier, and should be preferred.[15]

This alone, represents a powerful argument to eat a plant-based diet.

Now, allow me to rank the top dozen anti-aging, anti-cancer, antioxidant Superfoods, based on the latest findings by serving to make it more practical in terms of helping you find ways to eat healthier on a day-to day basis. The average antioxidant content of much of what Americans eat... peas, carrots, corn, lettuce, bananas just as a reference.

Let's go in reverse order for the "Top Dozen Superfoods" starting with number 12.

- In 12th place a tablespoon of **cocoa powder**: 400 antioxidant power (AP)
- In 11th place a half cup of **blueberries**: 4,047AP
- In 10th place is a **pomegranate**: 4,950AP
- In 9th place comes a **black plum**: 5,003AP
- In 8th place is a handful of **pecans**: 5,083AP
- In 7th place comes 1 **pear**: 5,233AP
- In 6th comes a half-cup of **cranberries**: 5,271AP
- In 5th place comes an **apple**: 5,609AP
- In 4th comes a teaspoon of **cinnamon**: 6,960AP
- 3rd comes an **artichoke**: 11,300AP
- And the runners up..... A half a cup of **goji berries**: 18,789AP

And if you have been given to choice to eat a single serving of food on the planet, in terms of antioxidant power, you would not have to go to the Himalayas for Goji's, but to Brazil. Way off the charts comes **Acai berries**.[16] Nearly 75,000 antioxidant units.

But, plums, pears and apples. You do not have to go to the other-side of the world or order exotic fruit over the internet. We hear how amazing pomegranates are, but one plum, one pear, one single humble apple, which brings up an important point....cost.

You can pretty much walk into any big natural food store, go to the food section and buy a pound of Aaci berries, but they may cost $5 or $10. So, in terms of practicality, we can calculate not only the antioxidant power of per serving, but also the important antioxidant content of these foods per dollar.

Superfood Bargains

Here we go.... Pecans – more than 8,000 antioxidant units per $. Great bargain. But, apples are even better. Goji berries are packed with antioxidants, but they are so expensive, that for the same price you can get abundantly more antioxidant power by eating cranberries or artichokes. Here is where Acai berries come in....so you are staring at that $5 package of frozen Acai pulp at the store and think about all the great smoothies you can make out of it. But, should you choose something a little more economical? Well, $ to $ Acai is worth it if you are willing to buy an apple a day to keep the doctor away. Better to go with Acai berries than apples. Why? Because Acai berries, are indeed, five to ten times more expensive per pound of apple, but with 20 times more antioxidants, so it makes sense to choose Acai berries vs. apples.

But, the 3 best bargain foods, that win the bronze, silver and gold prize for their antioxidant content are:

- Bronze: **Cloves**
- Silver: **Cinnamon**
- The Gold... busts off the charts as the No.1 antioxidant bargain in the world... **Red** or **Purple Cabbage**.

The cheapest food you can get and it is packed with antioxidants. And it last forever. So, next time you go shopping, buy a **red cabbage**. Put it in a crisper drawer in your refrigerator. Slice off shreds to put in as many things as you can think of. A great crunch for salads. Good for adding to soups, stir-fry's, whatever. In terms of eating healthy on a budget, purple or red cabbage *cannot* be beat by any food anywhere.[17]

The healthiest foods by weight only, without a doubt, are "Herbs and Spices." The best herbs & spices are, **Cloves and Oregano, All Spice, Cinnamon, Thyme, Sage and Peppermint**.[18]

What is the Minimum Recommended Daily Allowance of Antioxidants?

We must consider that the word "carbohydrates" means basically "hydrated carbon", that is, carbon dioxide + water, which is what plants utilize to make carbohydrates, and that is what is left when we burn them for energy: that we need to power our muscles, brain and other cells and tissues.

However, this process of oxidizing carbohydrates to make energy is complicated, messy, and generates free radicals. So, if drinking plain sugar-water causes the levels of oxidation in our bloodstream to increase over the following few hours,[19] why would the human body have a negative reaction to our primary fuel (carbohydrates)?

It is because during millions of years of evolution, there never was such a thing as "sugar water". All sugars and starches come pre-packed with powerful antioxidants!

In nature, sugar always comes with phytonutrients. If you drink the equivalent amount of sugar in the form of orange juice, you do not get that identical spike in oxidation. Why? It is because the sugar in fruits is pre-packed with phytonutrients. Cannot we, then, drink Vitamin C-enriched sugar water? No! It is not just the Vitamin C in the orange juice that reduces oxidative stress and mop-ups free radical damage in the body, but the citrus phytonutrients (more than 60 flavonoids, hesperidin, anthocyanins and a variety of polyphenols).

If we do not eat phytonutrient-rich foods with every meal, then for hours after we eat our bodies will be tipped out of balance in a pro-oxidative state. This could set us up for oxidative-stress diseases. For example, the free radicals in our body can increase the oxidation of the fats in our blood and can set us up for heart disease.[20] If we do not eat phytonutrient-rich foods with our meals, our body will have to dip into its reserves or back-up supply of antioxidants, and we cannot get away with that for long before disease develops.

Ideally, we *should* be consuming as many antioxidant-rich foods as possible. At the very least we should try to eat enough antioxidants to try to counter the oxidation of digestion.[21] This is how many antioxidants we need every day (depending on how much we eat) to just counter-act the effects of oxidation of digestion:

- Men in the U.S. average about 2,500 calories per day, so they ought to be getting at least 11,000 micro moles per day.
- Women who eat 1,800 calories on average per day should be getting 8,000 micro moles per day just to stay solvent. The average American does not even get *half* the minimum.

No wonder oxidation, stress-related, diseases abound. We are getting so few antioxidants from our diet that we cannot even keep up with the free radicals created by digesting our meals. The American nation is in chronic oxidized debt. In developed societies a lot of food is consumed, but not enough plants. This could result in exaggerated and prolonged metabolic, oxidative and immune imbalance, presenting an opportunity for biological insult that, over time, could overcome human defenses and repair systems, manifesting themselves in cellular dysfunction, disease and, ultimately, death.[22]

How to Reach the Human Antioxidant Recommended Daily Allowance?

To obtain our daily minimum of 8,000-11,000 antioxidant units per day all we have to do is to just eat lots of fruits and vegetables. Let us say, I eat a whole banana for breakfast in addition to whatever else I eat. Lunch will include the typical American salad with iceberg lettuce, some cucumber slices, and canned peaches for dessert. Supper will include a side serving of peas and carrots and, maybe, another salad, and I finish off with a slice of watermelon for dessert. This amounts to 9 servings of fruits and vegetables, and I may feel good about myself. But this adds up to only around 2,700, that is, less than a quarter of my minimum daily recommended antioxidant intake. What am I supposed to do, then? Eat 36 servings of fruits and vegetables per day? No!

What if instead of consuming that banana, I ate a single serving of blueberries for breakfast? Imagine that for lunch I have a salad that contains red cabbage, kale, onions, garlic, tomatoes, topped up with some kidney beans with spices. Then, I ate apples and dates for a quick snack. It is not even supper time, and I already will have had my 8,000-11,000 RDA of antioxidants! That is exactly why, it is *not* only the quantity, but the quality, of fruits and veggies that matters.[23] We should try to choose the healthiest plant foods that contain the most antioxidants.

But, if we do that, can we then skip fruits and vegetables for supper because we already have the RDA of antioxidants? Not a good idea. The estimated minimum antioxidant requirement does not account for the body's needs if other oxidant stressors – such as dietary pro-oxidants (meat), disease situations, exercising, cigarette smoke, air pollution, drugs, etc., are present. In that case we would need to consume a lot more plant antioxidants than the minimum level.[24]

Antioxidant levels can actually plummet within two hours of a stressful event, while subsequent recovery usually takes 3 days.[25] The take-home message is that it is especially important, whenever we are stressed, sick or tired, to go *above* the RDA of 8,000-11,000 antioxidant requirements. This produces the ideal situation of having our bloodstream soaking in antioxidants. This means consuming high-powered fruits and vegetables at every meal, berries or beans, and drinking hibiscus tea all day long.

In other words, select and eat the most powerful, most protective and most nutrient-dense, antioxidant plant-based foods, and you will be rewarded with good health and longevity.[26]

How about mushrooms? Mushrooms actually belong in the fungi group and are not considered real vegetables. I would highly recommend you all to eat the common "white-button mushroom". Diets high in mushrooms may modulate the aromatase activity and function in chemoprevention in postmenopausal women by reducing the in-situ production of oestrogen.[27] Later, we *shall* discuss more in detail, about the health benefits of mushrooms and their important role in breast cancer prevention.

Notes for chapter 16

1   Rui Hai Liu. (2003). Health benefits of fruits and vegetables are additive and synergistic combinations of phytochemicals. *American Journal of Clinical Nutrition*, **78**(3), pp.517S-520S; Temple, N.J. (2000). Antioxidants and disease: more questions than answers. *Nutrition Res*, **20**, pp.449–59; Willett, W.C. (1994). Diet and health: what should we eat? *Science*, **254**, pp.532–37; Willett, W.C. (1995). Diet, nutrition, and avoidable cancer. *Environ Health Perspective*, **103**(8), pp.165–70; Doll, R. *et al.* (1981). Avoidable risks of cancer in the United States. *Journal of the National Cancer Institute*, **66**, pp.1197–265; National Research Council. (1989). Diet and health: implications for reducing chronic disease risk. Washington, DC: National Academy Press; Dragsted, L.O. et al. (1993). Cancer-protective factors in fruits and vegetables: biochemical and biological background. *Pharmacology Toxicology*, **72**, pp.116–35; Clemens, M.R. *et al.*(1998). Effect of dietary phytochemicals on cancer development. *International Journal of Molecular Medicine*, **1**, pp.747–53.

2   Liska, J. A. (1998). The Detoxifcation Enzyme Systems. *Altern Med Rev*, **3**(3), pp-187-198. Silver, D. (2012). The Role of Detoxification in the Prevention of Chronic Degenerative Diseases. Retreived from http://www.deansilvermd.com/wp-content/uploads/2014/08/Detoxification-Research-Review-MET1836.pdf

3   Colleen Patrick-Goudreau. (Nov 1st 2010). *Color Me Vegan: Maximize Your Nutrient Intake and Optimize Your Health by Eating Antioxidant-Rich, Fiber-Packed, Color-Intense Meals That Taste Great.* Published by Fair Winds Press.

4   Campbell, T.C., & Campbell, T.M. (2006). *The China Study: Part 1:The China Study, Chapter 4:Lessons From China*, (1st ed), pp.92-93. Published by BenBella Books.

5   Ahmed M, Khan MI, Khan MR, Muhammad N, Khan AU, *et al.* (2013) Role of Medicinal Plants in Oxidative Stress and Cancer. **2**(2), pp.1-3. doi:10.4172/scientificreports.641

6   Ornish, D. (13th May, 2009). Mostly Plants. *American Journal of Cardiology*, pp.957-958. Published by Elsevier Inc. doi:10.1016/j.amjcard.2009.05.031. Ornish, D. (2008). *The Spectrum: A Scientifically Proven program to Feel Better, Live Longer, Lose Weight, and gain health.* p.386. Published by Ballantine Books.

7   Ibid.

8   Ibid.

9   Druesne-Pecollo, N *et al.* (2010). Beta-carotene supplementation and cancer risk: a systematic review and metaanalysis of ramomized controlled trails. *Cancer*, **127**(1), pp172-184.

10   Chaoyang Li, MD. (Nov 2010). Serum {alpha}-Carotene Concentrations and Risk of Death Among US Adults. The Third National Health and Nutrition Examination Survey Follow-up Study. *Arch Intern Med.* DOI: 10.1001/archinternmed.2010.440.

11   Ornish, "Mostly Plants", 2009.

12   Greger, M. (13th, August 2011). *The Antioxidant Power of Plant Foods vs. Animal Foods.* Volume 5. Retreived from http://nutritionfacts.org/video/antioxidant-power-of-plant-foods-versus-animal-foods/

13   Carlsen, M.H., et al. (2010). The total antioxidant content of more than 3100 foods, beverages, spices, herbs and supplements used worldwide. *Journal of Nutrition,* **9**(3). doi:10.1186/1475-2891-9-3

14   Greger, "*The Antioxidant Power of Plant Foods vs. Animal Foods*", 2011.

15   Carlsen, "The total antioxidant content of more than 3100 foods", 2007.

16   Greger, M. (September 19th 2008). Antioxidant content of 300 foods. Volume 2 video. Retreived from http://nutritionfacts.org/video/antioxidant-content-of-300-foods-2/

17   Greger, M. (Sptember 22nd 2008). Superfood Bargain. Volume 2 video. Retreived from http://nutritionfacts.org/video/superfood-bargains-2/

18   Steinar Dragland., et al. (2003). Several Culinary and Medicinal Herbs Are Important Sources of Dietary Antioxidants. *The Journal of Nutrition,* **133**(5), pp.1286-1290.

19   Mohanty, P. et al. (2000). Glucose challenge stimulates reactive oxygen species (ROS) generation by leucocytes. *Journal of Endocrinology Metabolism,* **85**(8), pp.2970-2973.

20   Zilversmit, D.B. (September 1979). Atherogenesis: a postprandial Phenomenon. *Journal of Circulation,* **60**(3) pp.473-485.

21   Ronald, L. et al. (2007). Plasma antioxidant capacity changes following a meal as a measure of the ability of a food to Alter in vivo Antioxidant status. *Journal of the American College of Nutrition,* **26**(2), pp.170-181.

22   Greger, M. (December 11th, 2013). *Minimum Recommended Daily Allowance of Antioxidants.* Volume 16. Retreived from http://nutritionfacts.org/video/minimum-recommended-daily-allowance-of-antioxidants/

23   Ibid.

24   Ronald, "*Plasma antioxidant capacity*", 2007.

25   Darvin, M.E. et al. (2008). One-year study on the variation of carotenoid antioxidant substances in living human skin: influence of dietary supplementation and stress factors. *Journal of Biomedical Optics,* **13**(4), 044028. doi: 10.1117/1.2952076.

26 Steinmetz, K. A., & Potter, J. D. (1991) Vegetables, fruits, and cancer: *Epidemiology, Cancer Causes Control* **2**,pp.325-337; Block, G, *et al.* (1992) Fruit, Vegetables, and Cancer Prevention: A review of the epidemiological evidence. *Nutrition and Cancer,* **18**, pp.1-29; Giovanucci, E, *et al.* (1995) Intake of Carotenoids and retinol in relation to risk of prostate cancer. *J. Natl. Cancer Inst,* **87**, pp.1767-1776; Di Mascio, *et al.* (1989) Lycopene as the most efficient biological carotenoid single oxygen quencher. *Arch. Biochemistry of Biophysic,*. **274**, pp.532-538; Stahl, W., *et al.* (1998) Carotenoid mixtures protect multilamellar liposomes against oxidative damage: synergistic effects of lycopene and lutein. *FEBS Lett,* **427**, pp.305-308; Amir H, *et al.* (1999) Lycopene and 1,25-dihydroxyvitamin D3 cooperate in the inhibition of cell cycle progression and induction of differentiation in HL-60 leukemic cells. *Nutr. Cancer,* **33**, 105-112; Levy, J., *et al.* (2000) Lycopene interferes with cell cycle progression and insulin-like growth factor I signalling in mammary cancer cells. *Nutr. Cancer,* **36**, pp.101-111; Stahl, W., *et al.* (2000) Stimulation of gap junctional communication: comparison of acyclo-retinoic acid and lycopene. *Arch. Biochemistry Biophysics,* **373**, pp.271-274; Broekmans, W. M., *et al.* (2000) Fruits and vegetables increase plasma carotenoids and vitamins and decrease homocysteine in humans. *Journal of Nutrition,* **130**, pp.1578-1583; Heber, D. & Bowerman, S. (2001) *What Colour Is Your Diet?* Harper-Collins/Regan Books, New York, NY.

27 Baiba, J.B. *et al.* (2001). White Button Mushroom Phytochemicals Inhibit Aromatase Activity and Breast Cancer Cell Proliferation. *J. Nutr,* **131**, pp.3288-3293.

17

The Cancer Environment

Cancer affects all of us, whether you're a daughter, mother, sister, friend, co-worker, doctor, or a patient.

Jennifer Aniston
American Actress, filmmaker and businesswoman

The Environmental Conditions Leading to Cancer Growth

If you wanted to grow a colony of cancer cells in the human body, at the cellular level, what would be the perfect environment? The perfect environment would offer the following:

- High acidity
- High blood sugar (Glucose and Fructose)
- High levels of sex hormones and cholesterol levels, and finally
- Low micronutrient/oxygen levels.

What would be the perfect diet to create such an environment? Definitely it is the standard American and Westernized "rich" diet, which the vast majority of people are currently eating.

I am not talking about a junk-food diet. I am talking about the same diet most people think is healthy. And why shouldn't they? This hazardous diet is being promoted by government and university experts (under the influence of food lobbies pushing high-fat foods). This is the same diet that is fundamentally altering the biochemistry of the body, weakening the immune system, and creating a welcome environment that allows cancer cells to thrive.

However, by changing the biochemistry of the body, through diet and lifestyle modifications, we can still eliminate the conditions that allow cancer cells to exist. Once we do that we will not need drugs to wipe out cancer cells.

With a healthy immune system the body is able to wipe out cancer and remove slow-forming microscopic tumours from our bodies. If the body's immune system is compromised (weakened) it cannot carry-out its normal protective powers. "Any treatment that does not address the underlying causes for the breakdown of the immune system will be palliative at best, and life threatening at its worst."[1]

Scientific evidence on cancer has shown that cancer thrives in *anaerobic conditions*. Cancer also prefers an *acidic pH environment* added with a good energy source of *carbohydrates* (starches) mainly in the form of glucose. Cancer is a sugar feeder. Medical scientists call it an "obligate glucose metabolite."

Listed below are high-starch foods that energize cancer, and which indeed, are not healthy food choices for the management dietary treatment for cancer patients:

- White foods (bread, pasta, rice, potatoes).
- Whole-grain pastry flour.
- Packaged cold cereals.
- Commercial/sweetened fruit juices and beverages.
- Confectionary foods: chocolate bars, cookies, cakes, and sweeteners (like honey, table sugar, maple syrup).
- Foods that contain high-fructose corn syrup.

Of course you should consider not only the quality of your carbohydrate choices, but also the quality of your fat and protein. We will discuss these nutrient issues later on in this book.

We must ask ourselves: Is the food I am about to eat a whole, natural plant source of essential nutrients? Is the food packed with antioxidants, fibre and micronutrients, and foods which comprise protective non-nutrients named phytochemicals which are found only and abundantly in plants? Remember that most of the beneficial nutrients are lost in foods once they have been industrialized, heavily processed and refined.

Avoid Overly Processed Foods

These are foods that include refined sugars, bleached flour and extracted oils. Refined and processed foods can also come from plant sources, but they have been mainly stripped from most of the healthy nutrients they once naturally contained. And although some nutritionists recommend olive oil (never cook with olive oil as the oil turns to trans-fats), olive oil is the most concentrated form of **fat** on the planet.

Oil is 100 percent pure and made of fat. This means that in the case of olive oil, for instance, manufacturers have taken whole olives, squeezed out and chemically extracted the good parts (healthy fibre, vitamins, and minerals), and left you with little more than a concentrated dose of calories. Olive oil contains a lot of calories in just one teaspoon (120 calories). So, is it a much healthier choice to always use the whole plant as food, so that you always get more richer, denser and healthier food nutrients.

The Western Diet and Cancer Risk

The Western diet is a good example of this cancerous environment. Today, most Western societies are consuming more animal products, more processed and refined sugary foods and less plant vegetation, thus contributing to the said conditions (anaerobic conditions, acidic pH levels, and higher blood-glucose levels). These conditions actually create the right favorable conditions for the origin and progression of this disease.

Dr. Otto Warburg, a cancer biochemist, theorized what he thought was the real cause of cancer way back in 1923. For this he received the Nobel Prize for medicine in 1931. He investigated the metabolism of tumours and the respiration of cells, particularly in cancer cells. In his book, "The Metabolism of Tumours",[2] he demonstrated that all forms of cancers are characterized by two basic conditions:

Otto Warburg

"Every single person who has cancer has a pH that is too acidic"

Dr. Otto Warburg won the 1931 Nobel Prize for proving that cancer cannot survive in an alkaline, oxygen-rich aerobic surrounding, but likewise, thrives in an acidic, oxygen-poor anaerobic environment.

- Hypoxia (lack of oxygen); and
- Acidosis (a pH value below pH 7.35).

Acidosis and Hypoxia are two sides of the puzzle: where you have one you most certainly have the other.

Any chemical, drug, toxin, carcinogen, food source or other damaging human cell stimuli agent, which renders the body more prone to an acidic environment, reduces the oxygen content of cells within the body, and interferes by altering human DNA may initiate the development of cancer. If the initiation factor is not removed the promotion and progression stage of cancer may prevail.

Dr. Warburg theorized that cancer occurs whenever any cell is denied 60 percent of its oxygen requirements, and showed that tumour cells exhibit anaerobic respiration. His thesis was that cancer is a fermentation disease caused by cells that have mutated from aerobic respiration to anaerobic respiration, resulting in glucose fermentation and uncontrolled cellular growth.

If there is inadequate amounts of oxygen for the cell to carry out sufficient aerobic respiration, cells stop breathing oxygen and start fermenting glucose (blood sugar) to create energy. The waste by-product of that fermentation process is a build-up of lactic acid.

Cancer utilizes glucose for energy; the by-product produces lactic acid, which brings on an anaerobic environment. The liver again metabolically turns this lactic acid back into glucose by a process called "gluconeogenesis" and this cycle continues to favour cancer cell growth.

In a 2011 documentary, "Cancer is Curable Now", Charlotte Gerson states:

> Once the body's pH goes into acidity, the blood becomes unable to transport sufficient oxygen, and the whole body's metabolism goes into a fermentative situation, instead of oxidative (normal energy production). The only thing tissues and cells can do with fermentative energy is grow and split, grow and split. And that is cancer! Virtually every patient who goes to a clinic with cancer has a pH below 7.0

(acidity). Within 1 week on the Gerson therapy the patient's pH value rises above pH 7.0 and the patient starts to heal, the tumour will start to shrink, and the patient's pain subsides.

Cancer Thrives on Glucose

A frequent characteristic of many malignant tumours is an increase in anaerobic glycolysis (the conversion of glucose to lactate) when compared to normal tissues.

Glucose is found in most foods, cancer metabolizes glucose into the by-product called lactic acid, which is subsequently excreted from the cancer cells into the blood. Blood carries the lactic acid to the liver, where it is later converted back into glucose to continue the cycle in feeding cancer. This occurs in all known cancers. It has been well documented and supported by clinical trials, as many years ago serum glucose levels were used to monitor the progress of cancer. It was concluded then that when the disease progressed, serum glucose levels automatically rise.

Positron Emission Tomography (PET) Scan is also used to establish if there is a high affinity or uptake of cells that require glucose. Patients are injected with fluorodeoxyglucose (FDG): a radioactive form of glucose. Then the patient is scanned, and whenever there is a rapid uptake of glucose, the oncologist can determine whether there is a metastatic focus that can determine if there is cancer, where it is developing, or if the cancer has metastasized to other parts of the body. The appropriate cancer treatment can then start immediately.

Malignant cells devour sugar faster than normal human cells. This is because *cancer cells have more glucose-receptor sites on the cell surface than regular human cells.* Cancer has a higher affinity for glucose (sugar) than normal human cells. The higher your blood sugar level is, the greater the probability that you are feeding cancer.

This is why a PET scan is a good diagnostic tool to detect whether there are any malignant cells present in the body. Cancer loves sugar very much.

The worst possible practical procedure method for treating cancer patients is to directly connect them to a glucose or dextrose intravenous drip. You would be feeding the cancer. Our objective should be to make it difficult for cancer cells to reproduce. So why feed them with a primary requirement?

Cancer cells find it hard to effectively use protein or complex carbohydrates for food, and so our body and the immune system will be able to use these energy sources as fuel and for repair mechanisms.

So, the patient should adopt a diet that includes plant proteins, plant sources of complex carbohydrates, plant fibre and essential fatty acids, and, at the same time avoid all the rest. In this way, the patient will also get an added bonus, as they will also be consuming the protective powers of the abundant vitamins, minerals and the non-nutrients compounds found in plants: the phytochemicals and antioxidants. This simple dietary change can make all the difference in the outcome of the disease process.

No Refined Foods

Most refined foods such as wheat flour, white flour, refined sugars, pastries, white pasta, white semolina, white bread, white rice (all food that is white, for that matter), French fries, chips of any variety, cakes, soft drinks, and similar junk foods and foods with a high content of fructose-corn syrup contain high amounts of carbohydrates. I think you know the foods I am talking about: foods that have been transformed from their natural state into man-made products.

Many types of refined foods contain no fibre. They spike your blood sugar, can cause insulin surges, stimulate your appetite, accelerate the conversion of calories into body fat and promote many diseases. From the body's perspective, eating these types of foods is, for all practical purposes, is the same thing as eating refined sugar. Refined foods do not fill you up, so you will be hungry shortly after eating them, which means you will have to eat even more calories to feel full. *Refined foods are the body's deadliest providers of energy sources for cancer.*

Many people think they are eating a plant-based diet because they are eating foods that do not contain animal products. But refined foods are every bit as harmful as animal foods, and should be eliminated from your diet. A plant-based diet is about eating natural, whole plant foods. The health benefits of this diet will be greatly diminished if refined foods are not eliminated.

So it is vital to stop eating refined foods and adding table sugar to your foods. Eat very little refined foods that contain added amounts of sugars. Many types of plant foods contain high amounts of natural or refined sugars. The ones to avoid are those that have been refined or contain high natural amounts of sugars: bananas, white bread, pasta, rice, potatoes, breakfast cereals, sweets, cookies, chocolate bars, refined fruit juices, soda drinks, and alcoholic beverages. Strawberries, peaches and all types of melons are sweet fruits, and should be avoided. You can tell they are sweet by simply tasting them.

Yes, fruits have marvellous cancer-fighting nutrients, but vegetables, nuts, seeds and legumes, non-sweet fruits and whole grains have too. Try to limit or eliminate sweetened fruits, and get all the cancer-fighting nutrients you need from other healthy anti-cancer plant foods. The diet described in this book has added the plant foods that contain low-glucose/fructose refined foods and sugary fruits, especially for this purpose. We do not want to feed the cancer. The goal is to starve cancer cells of their main supply of energy (glucose/fructose) as much as possible. Try to keep your blood sugar levels within the optimum range and avoid blood sugar surges, by consuming foods that have an "Low Gylcemic Index" (GI), that will not feed cancer. Obviously, once the cancer is under control, you can slowly and carefully reintroduce sweet fruits back into your daily-dietary regime.

Recommendations of Low GI Plant Foods for Cancer Treatment

Extremely healthy Low GI plant-based foods should include:

- Kale, broccoli, Brussels sprouts, watercress, collard greens, bok choy, arugula, spinach, and other vegetables
- Herbs and spices like turmeric, cloves, cinnamon, ginger, garlic,

oregano, and other types of vegetables like onions, leeks, red and white cabbage, cauliflower, artichokes, carrots, beets, including legumes (beans and lentils)
- Fruits like lemons, limes, grapefruit, cucumbers, orka, tomatoes, bell peppers, pumpkin, avocados, eggplant and apples.
- Nuts and seeds like walnuts, Brazilian nuts, almonds, flaxseeds, apricot seeds, sesame seeds, hemp and chia seeds.

There are many more plant foods to choose from. However, the above list of plant foods are excellent guidelines of plant-derivative Low GI plant food choices. These foods furthermore contain high amounts of dietary fibre, healthy plant-based protein, and essential fatty acids, not forgetting that, of course, they are also packed with essential vitamins, minerals and anti-cancer phytonutrients.

The Correct pH for Normal Human Cells and Tissues – an Alkaline pH

Blood must always maintain a pH of approximately 7.35-7.45 so that it can continue to transport oxygen. The body has the ability to be self-correcting by a system we call the "buffer system." A buffer is a substance which neutralizes acids, thus keeping the pH of a solution relatively constant despite considerable amounts of acids and bases. However, because of most people's poor diets of processed foods, junk foods, animal products (acidic-forming foods) and refined sugars, most people find it hard to balance their pH levels to the approximately 7.35-7.45 safety level.

Other factors are involved, but here we are talking about food, and what causes the development of cancer. So, although our bodies can counter-balance this pH problem and has many ways in maintaining typical alkaline reserves (which are utilized to buffer acids in these types of situations) it is safe to say that many of us have depleted these reserves – by the acidic-forming foods we consume.

In other words, once the buffer system reaches overload and cannot handle the additional acidic load from the acidic-forming foods we consume, the remaining acids are retained in our cells and tissues. As more and more acid is accumulated, our cells/tissues begin to lose their normal ability to function and may begin to deteriorate, mutate or die.

Yes! There are circumstances where damaging stimuli can make human cells and tissues try to compensate for these different environmental changes caused by toxic build-up, oxygen depletion (ischemia), DNA damage and other factors that can alter the standard biological process of ordinary cell development. Certain long-standing environmental stimuli can render human body cells internally

unsuitable for some specialized cell types and, as an adaptive response, the proliferating cells change their pattern of proliferation and differentiation.

Human cells have the remarkable adaptation ability to transform into a newer, more mature, and more stable type of cell, which better equips them to tolerate these environmental stresses. This process is termed "metaplasia". However, if the toxic overload is too great for the cell, then either the cell will terminate itself, a medical term called "apoptosis" (programmed cell-death) or stimuli will repair key cellular systems, causing dysfunction outside an adaptable range, after which another type of cell death may occur, a process termed "necrosis".

The other alternative possibility is that normal human cells may proliferate, differentiate and finally initiate into abnormal malignant cells.

As already mentioned, animal foods and other industrial chemical toxins are initiating or promoting agents. Carcinogenic agents may cause genetic abnormalities and may render human cells susceptible to the initiation stage of disease. Promoting agents can also expose normal cells to damage and may cause the development of cancer, unless of course the prolonged exposure to that cancer-causing agent is controlled or eliminated from the body. If not controlled, then the promotion stage may lead to the progression stage of cancer, which may further lead to the spread of the tumour to other organs within the human body.

This may also happen when our body's oxygen is depleted or our pH balance is altered. We already know that cancer thrives in an anaerobic and acidic environment, plus it feeds and ferments glucose for its primary source of energy.

Whenever cells become acidic, less oxygen is absorbed, and cells begin to ferment glucose in order to survive. This concept is extremely essential for all to understand because cancer cells certainly like to thrive in an acidic, anaerobic environment and do not develop well in aerobic and an alkaline environment.

In his study on the prevention of cancers, Otto Warburg noticed that "When aerobic respiration functions normally, and is intact, no cancer can exist. To prevent cancer it is therefore proposed first to keep the speed of the blood stream so high that the venous blood still contains sufficient oxygen."[3]

Cancer, more than all other diseases, has countless secondary causes. Warburg notes that however for cancer, there is only one prime cause:

> ...The replacement of the respiration of oxygen in ordinary body cells by the fermentation of glucose. Human body cells meet their energy needs by respiration of oxygen, whereas cancer cells meet their energy needs in great part from fermentation. All normal body cells are thus obligate aerobics, whereas all types of cancer cells are partial anaerobes. Basically, they thrive on lactic acid and glucose and not on oxygen.[4]

Since we are beginning to understand what internal conditions actually make cancer thrive (an acidic pH and hypoxia) then it stands to reason that the opposite conditions (an alkaline pH and adequate oxygen levels) should make cancer cells inert or harmless. We must also keep in mind that cancer ferments a vast amount of glucose (sugar) to produce enough efficient energy needed for its growth.

Human aerobic respiration is extremely efficient as it can generate as many as 38 ATP molecules from each glucose molecule, while anaerobic respiration produces only 2 ATP molecules. Thus, anaerobic respiration (cancer) releases appropriately 1/19th of the available energy. So if we do the maths, we calculate that in order for tumour cells to obtain the same amount of energy as normal human aerobic respiration, they must metabolize at least 19 times more glucose. This is why we sometimes encounter the phrase: "**Cancer loves sugar**".

To make matters worse, according to Dr. Steven Ayre "cancer cells get their energy also by secreting their own Insulin-Like Growth Factor

(IGF) and Insulin, which helps the stimulation of rapid growth. This, furthermore aids cancer cells gobble up glucose at a faster rate". These are their mechanisms of malignancy. Insulin and IGF work via attaching to specific cell membrane receptors, and these receptors are 16 times more concentrated on cancer cell membranes (cell walls) than on normal human cells. These receptors are the key to Insulin Potentiation Therapy (I.P.T) which I will discuss in great detail below.

If we use insulin in I.P.T., the result is that the low dose of chemotherapy gets channelled specifically inside the cancer cells, killing them more efficiently, vastly lessening any chemotherapeutic harmful side-effects. I.P.T. is ingenious: it kills cancer cells by activating the very same mechanisms that cancer cells use to kill people.[5]

For this very reason, it is vitally important that we mention this relative type of cancer therapy: Insulin Potentiation Therapy (I.P.T).

I.P.T. has been in existence since 1930. It has also been successfully used since January, 1946. The cancer therapy basically consists of *using extremely low doses of chemotherapy.* By targeting a minute dose of chemotherapy (less than 1/10th of the typical conventional dosage) to the cancer cells – I.P.T. enhances toxicity to the cancer, whilst reducing toxicity to the patient. It is extremely effective, relatively inexpensive, and a safer cancer therapy that has been used successfully for over 60 years.

How does IPT work against cancer? As you may already know, insulin is a hormone used to treat diabetes. Insulin is the main hormone that can reduce the amount of sugar in the blood, so it is used for patients who are diabetic, especially for patients with Type I Diabetes, medically called Diabetes Mellitus.

Insulin manages the delivery of glucose across the cell membranes of cells, joining up with specific-insulin receptors scattered on the outer surface of cell membranes. Every cell in the human body has between 100 and 100,000 insulin receptor sites. The binding of insulin to these receptors allows sugar (glucose) and other substances to be transported inside. That is why diabetics, who are unable to produce adequate amounts of insulin, cannot admit sugar into their cells, with the result that high amounts of sugar are retained in the blood, causing hyperglycaemia (high blood sugar).

But, what does this have to do with cancer? Remember that *cancer loves sugar* and cancer cells are anaerobic. Hence, they produce energy by fermenting glucose, an extremely inefficient way to produce energy, and also one of the reasons why cancer patients lose weight. The cancer cells require such vast amounts of glucose (19 times more than healthy normal cells), that they literally steal it away from the body's normal cells, and starve the cancer patient in the process.

How does IPT work in killing cancer cells? Patients are told to fast for 6-12 hours, and then they get a minute, calculated dose of insulin. This tiny dose of insulin opens the cell membranes and induces hypoglycaemia (low blood sugar), making the patient weak and slightly dizzy.

As Dr. Ayre stated, cancer cells have around 16 times insulin and IGF-1 receptors: more than normal-human cells. By inducing hypoglycaemia, this ingenious mechanism causes cancer cells to open their receptors at a rate of 16 to 1, thus allowing the treatment to selectively-target cancer cells. It typically takes 30 minutes to induce hypoglycaemia. Then, 18-Fluoro-deoxyglucose (a radioactive glucose) is administrated.

Malignant cells will "think" that they are going to be fed some sugar, so they open up their cell membranes. At this point, the medical team pulls the "bait and switch" on the cancer cells: low doses of traditional chemotherapy is then administered intravenously to the cancer patient. The cancer cells gobble up the chemotherapy, "thinking" that this is just their normal renewed energy daily source of glucose, and are killed instantly by chemotherapeutic doses that are much lower than in typical-standard chemotherapy.

Healthy tissue cannot readily absorb the insulin, and patients require only 10 to 25 percent of a standard dose of chemotherapy.

Dr. Thomas Lodi is another naturopathic doctor; however, uses I.P.T. combined with other effective-cancer treatments. He has successfully won the battle against cancer for many of his patients. He says:

I.P.T. is nature's "bow" that allows us to aim straight into the large (cancer cells). It is gentler, so patients generally have about 5 percent of the side-effects. Patients do not lose their hair; do not experience nausea or organ damage. (...) Stages 1, 2 and 3 cancers are typically eradicated. But, as always, there is a challenge in keeping it gone, which requires good nutrition, sleep and stress reduction. (...) I.P.T. almost always diminishes the size and the numbers of metastases, therefore patients are relieved of many symptoms associated with their disease. They have increased energy, decreased pain, improved appetite, and other physiological functions are restored.[6]

Dr. Thomas Lodi's cancer care program contains three pillars:

1. Teach patients how to stop making cancer,
2. Target and eliminate cancer without harming the patient, and
3. Stimulate, rebalance and enhance the Immune System.

If just eliminating cancer were the solution, then the cure rate for cancer would be much higher. Dr. Lodi believes that a truly comprehensive cancer-treatment program "should not", address the question of how to keep the cancer from not coming back; the main goal is to prevent cancer in the first place.[7]

My own research and the combined results of studies undertaken by many experienced medical professionals, has finally lead me to what I theorize to be the cause of cancer. In a few words, this is what I believe:

Here are the major conditions that cancer cells thrive on:

1) Poor circulation and low nutrient/oxygen levels.
2) Toxemia.
3) High acidity.
4) High animal proteins, saturated fat and cholesterol levels.
5) High sex-growth hormones, especially IGF-1 and Insulin.
6) High blood sugar levels, especially glucose and fructose.
7) Lack of daily exercise.
8) Stress-hormonal responses.

All acid-forming, processed, and refined foods, especially meat and dairy produce, should be eliminated from our daily diet. Other contributing environmental factors and household and beauty products too, can be toxic and acid-forming. Included in this category are: pharmaceutical drugs, medications, household cleaners, skin-hands-face cosmetics, fluoride toothpastes, sunscreen lotions, antiperspirants, etc.

Nuts and seeds are actually slightly acidic; nevertheless, they do contain essential fatty acids, fibre and healthy micronutrients and phytochemicals. We should make sure that we combine them with alkaline-forming food groups, and try to consume them on a daily basis.

We should always add nuts and seeds to our vegetable salads. The unsaturated and essential fatty acids (omega-3 fatty acids) found in most nuts and seeds, especially in flaxseeds and walnuts, help to absorb the phytonutrients contained in plant vegetables, making their plant nutrients more bio-available to the human body.

Now that we know which environmental conditions can make cancer cells thrive, then it is logical to conclude that the opposite conditions, should make cancer cells revert to being harmless. So, it makes perfect sense that we discontinue consuming foods that make our bodies' oxygen depleted, more acidic, and highly dependent on sugar.

People who consume huge amounts of soda (as well as coffee or alcohol) are typically highly acidic, and are "cancer magnets." A can of soda can reduce the function of the human immune system by 50 percent for a period of 6 hrs!

Which is the optimum way to keep human body cells and tissues pH within its optical range? The finest practical approach is to eat high amounts of alkaline-rich plant foods. Whenever our internal body environment is alkaline, cells and tissues will not be oxygen-depleted, and therefore cancer cannot exist; the body will start to repair itself.

In fact the most important step you can do to create an ideal acid-alkaline balance in your body is to take control of your diet by making "healthier plant food choices".

The general rule is to base your daily diet mainly on *80% alkaline-forming foods* and *20% acid-forming foods.* Basically, eat a *"whole-rich plant-based diet."* This will result in a natural tendency for the body's pH/oxygen balance to be favorable.[8]

The following chart shows which chemicals and foods are acid-forming and which are alkaline-forming.[9]

Acidic	Alkaline
• A few vegetables and fruits	• Almost all vegetables and fruits
• All animal products	• Raw nuts
• Whole and refined grains	• Sprouted grains and legumes
• Spices, herbs, condiments, spicy foods	• Wheat and barley Grass
• Sugar	• Sodium Bicarbonate
• Drugs, medications, tobacco	• Root vegetables
• Tea, coffee, soda drinks and alcohol	• Grounded flaxseeds
• Salt and fried foods	• Tofu
• Sports and energy drinks	
• Food additives & artificial sweeteners	

In earlier times, humans ate a diet consisting primarily of "whole natural foods". Today, unfortunately, whole plant foods constitute *only seven percent of our diet!* 42 percent of our calories comes from animal foods, and a whopping 51 percent of our calories comes from refined foods.

	Historical Percent of Calories	Current Percent of Calories
Animal Foods	5 %	42 %
Refined Foods	10 %	51 %
Whole Plant Foods	95 %	7 %

It is estimated that today a high percentage of hospital patients suffer from diet-induced or diet-associated diseases.[10] Is it any wonder, then, that not only cancer, but also other chronic-degenerative diseases are destroying our health?

Notes for chapter 17

1. Bollinger, Ty. (Jan 31st 2011). *The Cancer Truth*. Retreived from www.cancertruth.net.
2. Warburg, O. *et al*. (April 1926). The Metabolism of Tumours in the body. *The Journal of General Physiology*, pp.520-529.
3. Warburg, O.H. (February 24th 1956). On the Origin of Cancer Cells. *Science,* **123**(3191), pp.309-314. DOI: 10.1126/science.123.3191.309. Warburg, O.H. (June 30th 1966). The prime cause of cancer and prevention of cancer: respiration of oxygen in normal body cells vs. fermentation of sugars in cancer cells: The way to prevent cancer. (Revised Lecture at the meeting of the Nobel Laureates at Lindau, Lake Constance, and Germany).
4. Ibid.
5. Retreived from www.contemporarymedicine.net/pt_main.htm
6. Lord, T. An Oasis of Healing: IPT Low Dose Chemotherapy. Retreived from http://www.anoasisofhealing.com/our-program/target-eliminate-cancer/ipt-low-dose-chemotherapy/
7. Lori, T. Alternative Cancer Treatment Center: Science and Nature in Balance. Retreived from http://www.anoasisofhealing.com/
8. Ibid.
9. Brown, E.S. (2013). The Acid-Alkaline Food Guide: A Quick Reference to Foods & Their Effect on pH Levels (second ed), pp.79-164. Published by Square One.
10. Anderson, M. (2009). *Healing Cancer from Inside Out*. Part 2: Reversing Cancer: High Acidity, pp. 63-64. Published by www.RaveDiet.com . Esselstyn, C.B. as stated in the film *Eating* by M. Anderson; McDougall, J.A., The McDougall Program. p.17.

18

Supplementations, Juicing and Blending

Will we continue to do all we can to protect the public health against these dietary supplements that have been found to cause serious illness and injury?

Andrew von Eschenbach
Former Commissioner of the United States Food and Drug Administration from 2006–2009

Supplementations! "Yes or No?"

The answer is simple... No! Most vitamins and mineral supplements prescribed or given by your general practitioner for any medical condition – will not work as well as any natural nutrient component found in plant foods. Only in specific rare circumstances, such as when the patient has a genetic or acquired disease, then supplements may be given.

Vitamin B_{12} and D_3 are one of the few supplements which I would highly recommend, and are without any doubt, indispensable if you are a vegetarian, vegan, pregnant or nursing mother and if you reside in a country where you are not exposed to enough sunlight.

Lester Packer and Carol Colman highly recommend some essential antioxidants: that will aid and boost our liver-detoxification system, remove free radicals from our bodies, and boost the important liver-detoxifying antioxidant Glutathione, e.g. vitamin E and vitamin C, Lipoic Acid, Pycnogenol, and Coenzyme Q10.[1]

In most cases taking a single vitamin, mineral or antioxidant supplementation is a waste of money, as these do not work in synergy in the human body. On the contrary, different nutrients, like those found in plant foods, actually do.

No Single or Multiple Supplement Alone will Reduce Cancer Growth

Supplements alone will not make a dent against cancer. You have to change your overall diet and lifestyle. Supplements can help, but by taking them alone without changing your diet and lifestyle, will *not* "blow" the cancer candle out![2]

Remember: Cancer is not a deficiency disease of a single nutrient[3] and cannot be healed with the application of a single nutrient or even multiple nutrients, particularly in synthetic form. A vitamin's benefit, in other words, will become apparent only if one is not getting adequate amounts of it. Supplementations can have a place in the short-term healing of diseases, but in the long-term, they will cause problems.

We have nutritional deficiencies in the area of micronutrients, particularly those that fight cancer. However, our bodies work in symphony, and the absorption of micronutrients or antioxidants from whole plant foods, will orchestrate the healing of the body, in a way that single or multiple magic-bullet supplements cannot.[4]

According to Mile Andersom – "You *cannot* put nature in a pill"[5]

The examples about supplementations are really endless. But they all point to the same conclusion: that artificial synthetic supplementations throws the body off balance and can result in serious and unexpected side-effects. Natural foods in their natural packages keep the human body in balance. Many studies have proven that vitamins and minerals in the form of a pill, capsule, powder, liquid, chelation, etc.., actually cause more harm than good.[6]

There are natural therapies that use high doses of specific supplements for remarkable disease treatments (megavitamin therapy). It is worth mentioning that elevated doses of vitamin C (ascorbic acid) and vitamin B3 (niacin) have to be used intravenously for the treatment of cancer and for reducing depression and cholesterol levels in the blood – but in these cases one is advised to contact a highly qualified medical specialist in Orthomolecular Medicine.[7]

As already mentioned, synthetic drugs can also cause adverse side-effects. Drugs can sedate and tranquillize, and they can often produce the desirable symptomatic relief and sometimes they are even life-saving. But, drugs also produce other changes in the body, many of which are detrimental; drugs have created a generation of over-medicated, walking wounded people.

Drugs can, at best, be palliative – they will never be curative. Drug treatments are a major public health concern, killing hundreds of thousands of people each year worldwide.[8]

The Importance of Juicing and Blending

I truly believe that if you do add natural-plant juices to your life, certainly cleanse the body from harmful toxins, boost your immune system, enhance your physical performance, help lower your blood pressure and sleep better at night – have more energy and better health than you probably dreamed possible.

By drinking juice, you are eliminating a digestive process –extracting the liquid from the food fibre – and efficiently supplying the body with a cocktail of nutrients. The juicer separates the juice from the fibre, so that, what you drink is pulp-free and your body receives the maximum amounts of nutrients in minutes. This type of juice is completely different than bottled, canned, or concentrated juices sold in your local supermarket.

The so-called "Juiceman", Jay Kordich explains why juicing is so important[9]

1. First, it is absolutely free – which is important because nutrients lose a lot of value soon after juicing. (due to the breakdown of nutrients by oxidation)
2. Second, juice from the juicer is not pasteurized, which means "cooked" and so is bursting with living cells, which indeed, are so vital to good health.
3. Thirdly, and very important; fresh juice is absolutely pure, free of additives or any preservatives. If truly possible, do try to buy organic fruits and vegetables that have not been sprayed with any harmful pesticides. If you cannot buy organically grown foods, please, don't forget to thoroughly wash your fruits and vegetables before juicing.

There are many juicing devices and juicing mechanisms, but it all depends how much money you are willing to spend. If you have cancer, it would be reasonable to say to you – buy the most effective juicer on the market: A juicer that retains the highest percentage of juice, provides the most nutrients and removes the majority of pulp from your fruits and vegetables. The most common juicers are centrifugal juicers, which I would not recommend. The other types are masticating and Titurator (Grinder) Press combination juicers, which are far more superior than centrifugal juicers. These I would highly recommend if you have any form of chronic disease. The juicer I personally use and have in my own kitchen is a masticating juicer – the VitalMax Oscar 900.

Removing Pesticide Residues From Your Fruits & Vegetables

Agricultural pesticides cannot be removed with water alone (or the food industry would not use them). Luckily, just adding washing-up liquid (detergent) to water and generously swishing the fruit or vegetables around for a couple of minutes, can often lift off much of the pesticide residue. (You can test this by dipping organic grapes in water, and comparing this with dipping pesticide-laden grapes in water), and then in soapy water. The pesticide content is immediately obvious.[10]

Washing Fruit and Vegetables with Vinegar & Water

Some people recommend using vinegar; use one part vinegar to three parts water. This is great for reducing the bacterial load, and may also help break down the coating of wax found on some fruits and vegetables. The editors of *Cooks Illustrated* magazine tested this theory by using four different methods to clean pears and apples: a vinegar and water solution (3:1, water:vinegar), antibacterial liquid soap, scrubbing with a stiff brush, and just using plain water. Not only did the vinegar mixture work the best, it was far, far better when measured for bacteria — it removed 98% of bacteria, compared to just under 85% for scrubbing. The quickest way to do this at home is to keep a bottle of vinegar with a spray-top — just spray the fruit or vegetables with vinegar, then rinse under a tap. If you've got longer time to spare, leave fruit or vegetables soaking for 10-20 minutes in a vinegar/water solution, then rinse.

Discarding Outer Layers of Vegetables

Eat only the inner layers of food produce that you won't be cooking: such as, lettuce, cabbages, and other salad vegetables (including onions). Discard the outer layers, as these will have more pesticides resting on them from crop spraying. Assume that the outside layer of any fruit or vegetable will have absorbed most of the pesticides (though some will have also have been absorbed from the soil) so instead, wash/peel or discard these outer layers whenever possible.[11] Although washing with plain water can accomplish a lot, adding some natural sources of acid (namely lemon and vinegar) to the wash can provide a bit of additional, natural disinfecting power.

Follow these easy steps in making your own fruit and vegetable wash, that is both inexpensive and completely organic.[12]

1. Use an organic lemon. You *can* use a normal lemon, which would be slightly cheaper, but the wash couldn't be called "organic", just "natural". Regardless, both kinds of lemons will be fine for the task.
    - Slice the lemon in half.

- Squeeze out one tablespoon of lemon juice into a spray bottle. The lemon juice is both a natural disinfectant and will also leave your fruits and vegetables smelling nice.
2. Pour 2 tablespoons of vinegar in the spray bottle, along with one cup of water. The vinegar provides some additional disinfectant power.
3. Screw on the top. Shake the mixture vigorously.
4. Spray the wash on all of your fruits and vegetables; prior to using them. After each spray, I would recommend to leave this mixture on your foods for 30-60 mins, *before* rinsing, juicing and eating them.

According to the late Ann Wigmore who co-founded the Hippocrates Health Institute:

Ann Wigmore was an early pioneer in the use of wheatgrass juice and living foods for detoxifying and healing the body, mind, and spirit.[13] This is the woman who introduced wheatgrass to the world[14] and reversed her own cancer, using natural foods with supplements – after doctors had declared her terminally ill! Juicing was a big part of her nutrition, but later in her life she discontinued using it and turned to "blending", because juicing, "she said", is not the way food is found in nature.

Juicing can be very beneficial to the human body, but juicing vegetables and fruit have one big downside: it leaves behind most of the fibre in the food: Fibre is critical in the fight not only against cancer, but also against all other diseases. I refer using juicing as an optional supplementation. There are clinics which use *mainly* juicing as a treatment for cancer patients, and they have registered a significant success. I believe, however, that their success rates could have been much higher, if they used juicing as a supplement, not as the main source.[15]

The importance of Blending

Unlike juicing, blending includes the whole food; nothing is left behind. In fact, blending can be part of your every daily diet and is an

excellent way to get more green vegetables into your meals. Because green blending is very concentrated, you need to add small amounts of semi-sweet vegetables or fruits to them, such as, carrots, beets, apples, and tomatoes, in order to make them more palatable.[16] We need fibre – no one can live on juice alone – you can get fibre by eating around the juicer, by using a blender.

I would highly recommend juicing and blending for retaining most of your essential nutrients, including your daily-dietary fibre. You can accomplish this by using *both* juicing and blending in making nutritious and freshening juices, smoothies, soups and salads.

Notes for chapter 18

1. Packer, L. & Colman, C. (December 10th 1999). *The Antioxidant Miracle: Put Liopic acid, Pycnogenol, and vitamins E and C to Work for you* (1st ed). Published by Wiley.

2. Anderson, M. (2009). *Healing Cancer From Inside Out:* Part 3: The Rave diet and lifestyle: Supplementation, pp. 119-21. Published by www.RaveDiet.com

3. Proponents of laetrile will argue that cancer is caused by a deficiency in vitamin B_{17}, but B_{17} supplementation is neither the cause of nor the cure for cancer. Laetrile can help prevent cancer and reduce cancer reduction, but it is not the "miracle" vitamin for cancer cure and if you do not address the underlying cause of the tumour, it will return.

4. Anderson, "Healing Cancer from inside out", 2009.

5. Ibid.

6. L.A.Times (November 10th 2003). *Tomatoes may be better against cancer than Lycopene alone.* Galloe, A.M. (September 4th 1993). Influence of oral magnesium supplementation on cardiac events among survivors of an acute myocardial infarction, *BMJ*, **307**(6904), pp.585-7. Black, M.R. (June 1998). Zinc supplements and serum lipids in young adult white males. *AM J Clin Nutr,* **47**(6): pp. 970-975.

7. Hoffer, A. & Saul, A.W. (2008). *Orthomolecular Medicine For Everyone. Megavitamin Therapeutics for Families and Physicians.* Published by Basic Health Publications, Inc.

8. Ibid.

9. Kordich, J. (1992). *The Juiceman's Power of Juicing: Chapter 2: Why Juice?* (1st ed), pp.26-27. Published by HarperCollins, Canada.

10. Tomley. S. Mind and Body: How to Wash Pesticides Off Fruit & Vegetables. Retreived from https://suite.io/sarah-tomley/2nka2o6

11. The Trive. (Oct 10th 2013). How to Wash Pesticides Off Fruit & Vegetables. Retreived from http://healthyfoodtribe.com/how-to-wash-pesticides-off-fruit-vegetables/

12. Wiki How. How to Make a Organic Fruit and Vegetable Wash. Retreived from http://www.wikihow.com/Make-an-Organic-Fruit-and-Vegetable-Wash

13. NCAHF Newsletter. (Sep 1st 1994). Wheatgrass therapy

14. Wigmore, A. (Oct 1st 1985) *The Wheatgrass Book: How to Grow and Use Wheatgrass to maximize Your Health and Vitality* (1st ed). Published by Avery Trade.

15. In certain cases, where juicing is the only way to make someone intake nutrients, because of some medical condition, then juicing can have the main role.

16. Anderson, "Healing Cancer from inside out", 2009.

19

The China Study with Reference to Breast Cancer

Furthermore, a pattern was beginning to emerge: nutrients from animal-based foods increased tumour development, while nutrients from plant-based foods decreased tumour development.

T. Colin Campbell
The China Study: The Most Comprehensive Study of Nutrition Ever Conducted and the Startling Implications for Diet, Weight Loss and Long-Term Health

The universal best-selling book entitled "The China Study" was written by T. Colin Campbell and Thomas M. Campbell II. These two influential clinical scientists with the help of their team of colleagues performed the most comprehensive and largest study ever undertaken: on the relationship between diet and the risk of developing disease within the Chinese population.

T. Colin Campbell's laboratory work was mainly focused on several cancers: the human liver, pancreas and the breast, etc. We shall focus mainly on the impressive and vital data that Campbell gathered from China, relating to the most common cancer that affects mostly women in the Western world, namely breast cancer.[1]

What he found was that the nutritional effects on the cancers he studied were virtually the same for all cancers, regardless of whether they are initiated (started) by distinctive factors or whether they are located in different parts of the body. This clearly demonstrates that there is evidence linking food to many health concerns.

It is clear that breast cancer is an important concern in women in all societies, but especially in developing fast-food nations, that actually have the highest rates of breast cancer in the world.

This disease, perhaps more than any other, is a major health concern, and it causes a sense of panic and fear in women. Dr. Campbell noticed that there are at least four important breast cancer risk factors that can be affected by nutrition.

Campbell concluded that, with the exception of cholesterol, these risk factors are variations on the same theme:

- Exposure to excess amounts of female hormones, including estrogens and progesterone, leads to an increased risk of breast cancer.
- Women, who consume a diet rich in animal-based foods and low in plant-based foods, reach puberty earlier and menopause later, thus extending their reproductive life. These women have higher levels of female hormones throughout their lifespan, which can also lead to an increased risk of the disease.

Breast Cancer Risk Factors and Nutritional Influence

Risk of breast cancer increases when a woman has:	A diet of animal foods and refined carbohydrates:
Early age menarche (first menstruation)	Lowers the age of menarche
Late age of menopause	Raises the age of menopause
High levels of female hormones in the blood	Increases female hormone levels
High blood cholesterol	Increases blood cholesterol levels

Colin Campbell's data on The China Study shows that exposure to oestrogen is at least 2.5-3.0 times higher among Western women when compared with rural Chinese women – a huge difference for such a critically important hormone. Furthermore, there is overwhelming evidence that oestrogen levels are a fundamental determinant of breast-cancer risk. This is because increased levels of estrogens and relative hormones (testosterone and progesterone) are a result of the consumption of typical Western diets that have a preference for animal foods, that are:

↑High in fats   ↑High in animal protein   ↓Low in fibre[4]

Campbell recognized that the risk of breast cancer is preventable if people choose foods that keep oestrogen levels under control. The sad truth is that most women simply are not aware of this scientific medical evidence. If this vital information were properly reported by responsible and credible public health agencies, then many young women would be taking very real and effective steps to avoid this awful disease.

Environmental Chemicals

Many environmental chemicals have been shown to disrupt hormones, although it is not clear which hormones in humans are being disrupted. Commonly associated with industrial pollution, dioxins and PCBs, persist in the environment because they are not metabolized when consumed. Thus, they are not excreted from the body and may build-up in the fat and breast milk of lactating mothers. Some of these chemicals are known to promote the growth of cancer cells. Humans may not be at significant risk unless one consumes excessive amounts of meat, milk and fish. Indeed, 90-95% of our exposure to these chemicals come from consuming animal products – yet another reason why consuming animal-based foods can be risky.

There are more environmental chemicals that also commonly perceived to be significant causes of breast[5] and other cancers. They are called Polycyclic Aromatic Hydrocarbons (PAHs) and are found in auto exhaust, factory smoke stacks, petroleum, tar products and tobacco smoke, among other processes common to an industrial society. Fortunately, these types of chemicals humans can metabolize and excrete them. But there is a problem: when the PAHs are metabolized within the body, they produce intermediate products that react with human DNA to form tightly bound complexes. This is the first step in causing cancer. In fact, these chemicals have recently been shown to adversely affect the BRCA-1 and BRCA-2 genes of breast cancer cells grown in the laboratory.[6]

Hormone Replacement Therapy (HRT)

Most women today know about the side-effects and dangers involved in HRT, which can also increase breast cancer risks. HRT is still taken by many women: to alleviate the unpleasant effects of menopause, to protect bone health, and to prevent coronary heart disease.

However, it is now becoming widely acknowledged that HRT is not as beneficial as once thought and it can have certain severe side-effects. What are the facts?

In the last few decades, large trials of HRT have been conducted, particularly randomized intervention trials: such as, the Women's Health Initiative (WHI) and the Heart and oestrogen/progestin Replacement Study (HERS).

The WHI trial showed that after 5.2 years of taking HRT, women have a 26% increased risk of contracting breast-cancer, while the HERS study saw an even greater increase in risk: 30%. These two studies are consistent, and indicate that increased exposure to female hormones via HRT does indeed lead to more breast cancer incidences.

HRT is associated with lower rates of coronary heart disease, but this is not necessarily true. In the large WHI trial, for every 1,000 healthy postmenopausal women who took HRT, there were 7 more women with heart disease, 8 more with strokes, and 8 more with pulmonary embolism – the opposite of what had been expected. HRT can increase cardiovascular disease risks after all.

With all this data and by using simple maths, we can easily deduce that HRT could well be the cause of more harm than good.

Instead of relying on HRT, I think that there is a better solution. I personally would suggest using food! Plant foods will not only protect your bones, your heart, and help you relieve menopausal discomforts, but they will also reduce your risk of developing breast cancer.

Dr. T. Colin Campbell found that:

- During a woman's reproductive years, hormone levels are elevated, although the levels among women who eat plant-based diets are not as elevated.
- At the end of a woman's reproductive years, it is entirely natural for reproductive hormones of all women to drop to a low "base" level.
- As reproductive years come to an end, the lower hormone levels among plant-eaters do not crash as hard as they do among animal-eaters. These abrupt hormonal changes in a woman's body are what cause menopause symptoms.
- Therefore, a plant-based diet leads to less severe hormone crash and a gentler menopause.

Campbell concluded:

> Even if future studies fail to confirm these details, a plant-based diet still offers the lowest risk of both breast cancer and heart disease, and it might just be the best of both worlds, something that no drug can offer. (...) In each of the environmental issues involving breast cancer risk, the use of Tamoxifen, HRT environmental chemical exposure, and preventive mastectomy are distractions that do not allow us to consider a safer and far more useful nutritional strategy: a "healthy plant-based diet". It is critical that we change the way we think about this disease, and that we provide this vital information to all the women who need it.[18]

This is why the protective chemical compounds found abundantly in plant foods are our main defense, against these environmental pollutants and other detrimental causes. These environmental and dietary factors can increase a woman's and a small minority of men's risk for developing breast cancer.

Notes for chapter 19

1. Campbell, T.C. & Campbell, T.M. (2006). *The China Study: The Most Comprehensive Study of Nutrition Ever Conducted*. Part 2: Diseases of Afflence: Chapter 8: Common Cancers: Breat, prostate, Large Bowel (Colon and Rectal), pp.158-161. Published by BenBella Books.

2. Wu, A.H., *et al.* (1999). Meta-analysis: dietary fat intake, serum oestrogen levels, and the risk of breast cancer. *Journal of the National Cancer Institute*, **91**, pp.529-534.

3. Rosenthal, M.B *et al.* (1985). Effects of a high-complex-carbohydrate, low-fat, low-cholesterol diet on levels of serum lipids and estradiol. *American Journal of Medicine*, **78**, pp.23-27. Adlecreutz H. (1990). Western diet and Western diseases: some hormonal and biochemical mechanisms and associations. *Scandinavia Journal of Clinical Lab Investigations*, **50**(suppl.201), pp.3-23.

4. Rose, D.P., *et al.* (1997). Effects of diet supplementations with wheat bran on serum oestrogen levels in the follicular and luteal phases of the menstrual cycle. *Nutrition*, **13**, pp. 535-539; Tymchuk, C.N *et al.* (2000). Changes in sex hormone-binding globulin, insulin, and serum lipids in postmenopausal women on a low-fat, high fibre diet combined with exercise. *Nutrition and Cancer*, **38**, pp.158-162.

5. Ronai Z *et al.* (2004). "Contrasting incidence of ras mutations in rat mammary and mouse skin tumours induced bt anti benzol(c) phenanthrene-3,4-diol-1,2-epoxide." *Carcingenesis*, **15**, pp.2113-2116.

6. Jeffy, BD *et al.* (1999). "Inhibition of BRCA-1 expression by benzo(a)pyrene and diol epoxide." *Mol Carcinogenesis*, **26**, pp.100-118. Campbell, "The China Study, 2006, pp.165-166.

7. Wu, "*Meta-analysis*", 1999.

8. Campbell, "The China Study", 2006.

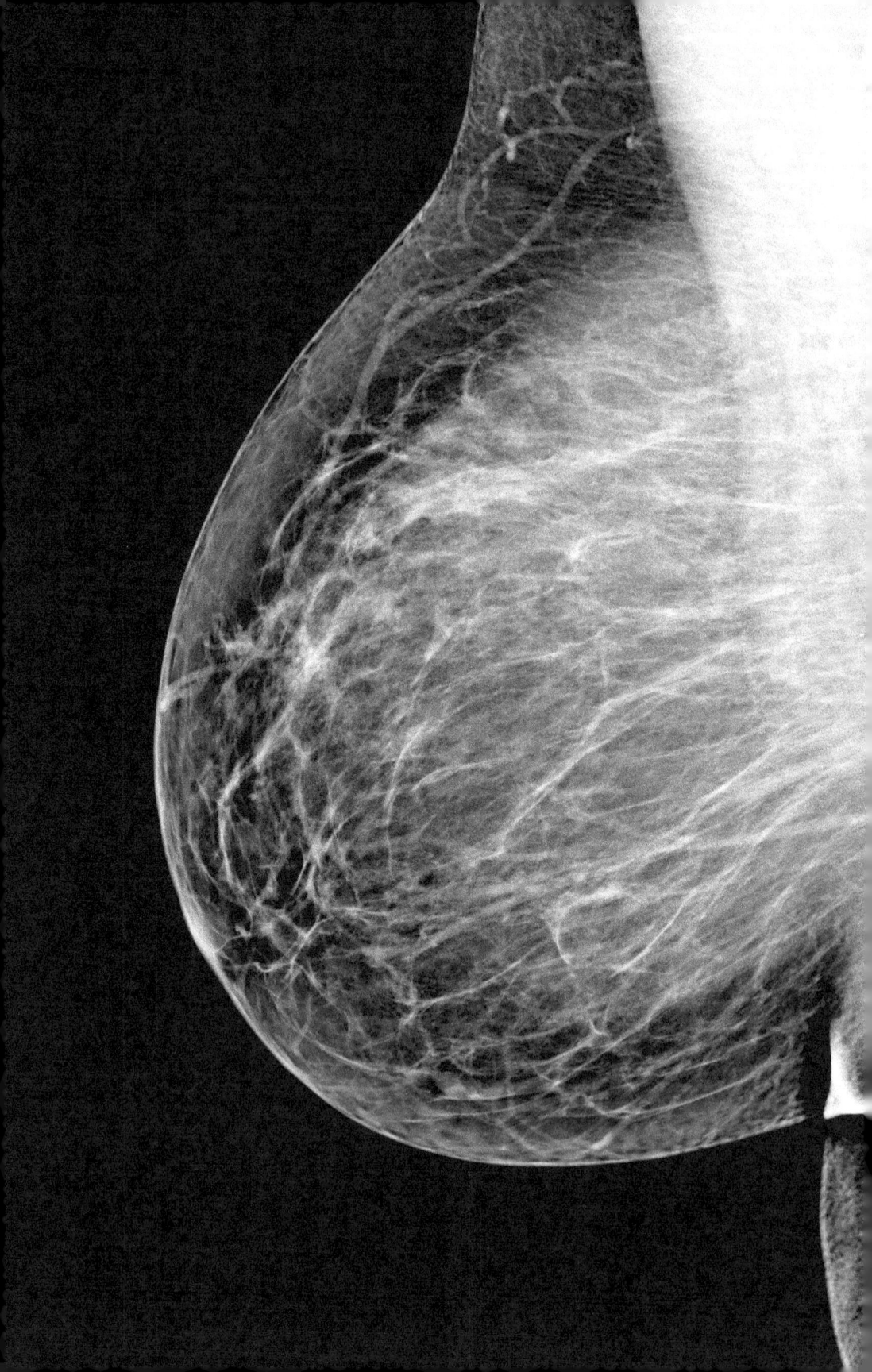

20

Mammography: Good or bad?

Mammogram interpretations are unreliable. According to an article in the Journal of the American Medical Association (May 26, 1993) one study revealed a false positive rate in the range of 20% to 63%. This suggests that huge numbers of women are unnecessarily going through the pain, expense, and anxiety of biopsies.

Michael Phillip Wright
*The Epidemic of Unnecessary Mastectomy:
How not to be a victim*

Mammography

Basically, a mammogram is an X-ray diagnostic instrument that takes a picture of a woman's breast and that can reveal tumour growths otherwise undetectable in a physical examination.

As with X-rays, mammograms use doses of ionizing radiation in creating an image. Radiologists can then examine the image of the breast for signs of any abnormalities. But, is a mammogram really an effective tool for detecting tumours? Many doctors are starting to say "No."

The Main Problem with Mammograms: Breast Density

"Breast density" refers to the relative amount of fat vs. connective tissue in epithelia tissues. That ratio could be primarily determined genetically in two-thirds of adult females in their forties who have dense-breast tissue. This is the problem with mammograms. Although breast density generally declines with age, up to a third of women retain dense-breast tissue for a number of years after menopause. So how do we know breasts are "dense"? You will need to read the details of your mammogram report. Radiologists classify breast density into four categories based on the appearance of the tissue on a mammogram:[1]

1. If the breast is < 25% dense, they call it fatty replaced.
2. If the breast is >25 but <50% dense, it is called scattered fiber granular density
3. If it is >51 but <75%, it is taken to be heterogeneously dense
4. And in the end, if the breast is >75% dense, it is regarded as extremely dense.

Breasts that fall into categories 3 and 4 are considered dense. The problem with breast density is that it is really the "beast in sheep's clothing". Both tumors and dense breast tissue appear "white" on a mammogram, and the X-ray often cannot differentiate between normal breast tissue and breast cancer. So, it is easy to distinguish a tumour in the upper part of a fatty breast, but it is extremely hard to see a tumour if the breast falls into category 3 or 4. That is why mammograms find over 80% of tumours in fatty breast, but as few as 40% in extremely dense breast tissue.[2]

In a Swedish study of 60,000 women, it was estimated that around 70% of the tumours detected by mammograms turned out to be false positives. These "false positives" are not only physical, emotional and financial strains on the victims, but they also lead to many unnecessary invasive biopsies.[3]

In "The Politics of Cancer" Dr. Samuel Epstein writes:

> Regular mammography of younger women increases their cancer risks. Documents in many control trials over the past decade have shown consistent increases in breast cancer mortality within a few years of commencing screening. The evidence corroborates that the premenopausal breast can be highly sensitive to the cumulative carcinogenic effects of irradiation.[4]

A mammogram should never be used for early detection of breast cancer, as it actually increases your chances to develop the disease. Ionizing radiation is a risk factor agent that could lead to cancer.

According to a 2000 article in the The Lancet, "Screening for breast cancer with mammography is unjustified (...) there is no true evidence that screening decreases breast cancer mortality."[5]

According to Mike Adams (the "Health Ranger") of Natural News:

A report released by the National Council on Radiation Protection and Measurement reveals that Americans' exposure to radiation has increased more than 600% over the last three decades. Most of that increase has come from patients' exposure to radiation through *medical imaging scans: such as CT scans and mammograms.*

Most patients have no cognizance of the dangers of ionizing radiation due to medical imaging scans. For example, virtually no patients and only a few doctors actually realize that one CT scan can expose the body to the equivalent of several hundred X-rays. Most women undergoing mammograms have no idea that the radiation emitted by mammography machines may actually cause cancer: exposing heart and breast tissue to dangerous ionizing radiation, that directly causes DNA damage.

Dr. John Gofman has made this startling statement about the impact of medical radiation:

"Our estimate is that about three-quarters of the current annual incidence of breast cancer... is being caused by ionizing radiation, primarily from medical sources."[6]

There is more about the unnecessary usage of Mammography. In the New England Journal of Medicine (February 11, 1993) said that:

"Of every 1,000 American women getting mammograms each year between the ages of 40 and 50, 345[35%] will receive false positive results, often with unnecessary intervention [i.e., treatment] as the result."

and "....the annual mammagraphic screening of 10,000 women aged 50-70 will extend the lives of, at best, 26 of them. Also, the annual screening of 10,000 women in their 40s will extend the lives of only 12 women per year." –How Mammography Causes Cancer," Alternative Medicine, Sep. 1999

Using sources of known data taken from 22 pedigree studies (studies done on the entire family, not just on a person) of 8,139 subjects, the research team estimated that for BRCA-1 or BRCA-2 mutation carriers, annual mammographic screening starting at 25 to 29 years of age would confer a lifetime risk of radiation-induced breast cancer mortality of 26 per 10,000 women.[7]

Mammography is the standard tool for determining the exact location of a developed tumour. Nevertheless, it can also increase a patient's risk of developing breast cancer. Many women also complain about the discomfort and the pain that it causes, as the pressure placed upon the breast during the procedure is nearly 50kg.

Nor is mammography an early-warning procedure, even though some women assume that it is. "Early" is a relative term, so if a mammogram can detect a tumour in the 8th year, it is earlier than the 10th year, but in any case, even the 7th year is too late to change the outcome. The real danger of breast cancer is whether or not it has spread to a vital organ. If it is going to spread, it has had many years to do so.

But, ideally and more respectively, our main challenge and focus should be on the earliest prevention possible, that is, preventing the emergence of the tumour in the first place and slowing its growth by using diet and lifestyle changes. Eating foods that prevent and inhibit the promotion and regression stages of cancer growth is the key to the battle against this disease. There is a healthy remedy and an early prevention strategy to breast cancer, and it is called a "Plant-Based Diet".

Numerous plant foods at present have been identified and studied, documented and reviewed in medical literatures, as having anti-cancerous properties. Numerous plant foods are proven to not only dramatically reduce your risk to develop breast cancer, but may also reverse it.[8]

We now know that a diet that excludes all types of animal-based foods is the best diet for breast cancer prevention and survival.[9]

It is hoped that mammography, chemotherapy, radiation and unnecessary surgery will no longer be applied for the purpose of detecting, diagnosing and treating breast cancer. It is hoped that many more people will finally be ready to "step outside the box" and realize that the best natural treatments and the real underlying solutions for the prevention, control, and long-term survival of this disease –is the usage of plant foods and lifestyle changes,[10] as explained throughout this book.

Notes for chapter 20

1. Deborah Rhodes. (2011). A tool that finds 3x more breast tumors, and why it is not available to you? TED Talks. TED stands for Technology, Entertainment, Design, and TEDTalks cover many topics, as well as science, business, development and the arts. Retreived from https://www.youtube.com/watch?v=DqbM1ZrpTQg

2. Ibid.

3. Lidbrink, E., et al. (February 3rd 1996). Neglected aspects of false positive findings of mammography in breast cancer screening: analysis of false positive cases from the Stockholm trial. *British Medical Journal, 312*(7026), pp.273-276.

4. Epstein, S. (1998). The Politics of Cancer, p.539. Published by East Ridge Press.

5. Gotzsche, CP et al. (April 7th 2012). Mammography Screening: Truth, Lies and Controversy. *The Lancet, 379*(9823), pp.1289-1290.

6. Gofman, WJ. (Feb 1996). Preventing Breast cancer: The Story of a Major, Proven Preventable Cause of This Disease. Published by Committee Nuclear Responsibility. See also, "The X-rays and Health project." An educational project of the Committee for Nuclear Responsibility. Retrieved from www.x-raysandhealth.org

7. Berrington de Gonzalez, A et al. (February 4th 2009). Estimated Risk of Radiation-Induced Breast Cancer from Mammographic Screening for Young BRCA Mutation Carriers. *Journal of the National Cancer Institute, 101*(3), pp.205-209.

8. Castello, A et al. (2014). Spanish Mediterranean diet and other dietary patterns and breast cancer risk: case-control EpiGEICAM study. *British Journal of Cancer, 111*, pp.1454-1462. doi:10.1038/bjc.2014.434. Donaldson, S.M. (2004). Nutrition and Cancer: A review of the evidence of an anti-cancer diet. *Journal of Nutrition, 3*(19) doi:10.1186/1475-2891-3-19. Rice, S and Whitehead, AS. (2006). Phytoestrogens and breast cancer –promoters or protectors? *Endocrine-Related Cancer, 13*, pp. 995–1015. Gaudet MM, Britton JA, Kabat GC, Steck-Scott S, Eng SM, Teitelbaum SL, et al. (2004). Fruits, vegetables, and micronutrients in relation to breast cancer modified by menopause and hormone receptor status. *Cancer Epidemiol Biomarkers Prev, 13*(9), pp.1485-1494. Riboli E, Norat T. (2003). Epidemiologic evidence of the protective effect of fruit and vegetables on cancer risk. *Am J Clin Nutr, 78*(3 Suppl), pp.559S-569S. La Vecchia C, Altieri A, Tavani A. (2001). Vegetables, fruit, antioxidants and cancer: a review of Italian studies. *Eur J Nutr, 40*(6), pp.261-267. Smith-Warner SA, Spiegelman D, Yaun SS, Adami HO, Beeson WL, van den Brandt PA, et al. (2001). Intake of fruits and vegetables and risk of breast cancer: a pooled analysis of cohort studies. *JAMA, 285*, pp.769-776.

9   de Lima FE, do Rosário Dias de Oliveira Latorre M, de Carvalho Costa MJ, Fisberg RM. (2008). Diet and cancer in Northeast Brazil: evaluation of eating habits and food group consumption in relation to breast cancer. *Cad Saude Publica,* **24**(4), pp.820-828. Wakai K, Tamakoshi K, Date C, Fukui M, Suzuki S, Lin Y, *et al.* (2005). Dietary intakes of fat and fatty acids and risk of breast cancer: a prospective study in Japan. *Cancer Sci,* **96**(9), pp.590-599. Taylor EF, Burley VJ, Greenwood DC, Cade JE. (2007). Meat consumption and risk of breast cancer in the UK Women's Cohort Study. *Br J Cancer,* **96**(7), pp.1139-1146.

10   McEligot AJ, Largent J, Ziogas A, Peel D, Anton-Culver H. (2006). Dietary fat, fiber, vegetable, and micronutrients are associated with overall survival in postmenopausal women diagnosed with breast cancer. *Nutr Cancer,* **55**(2), pp.132-140.

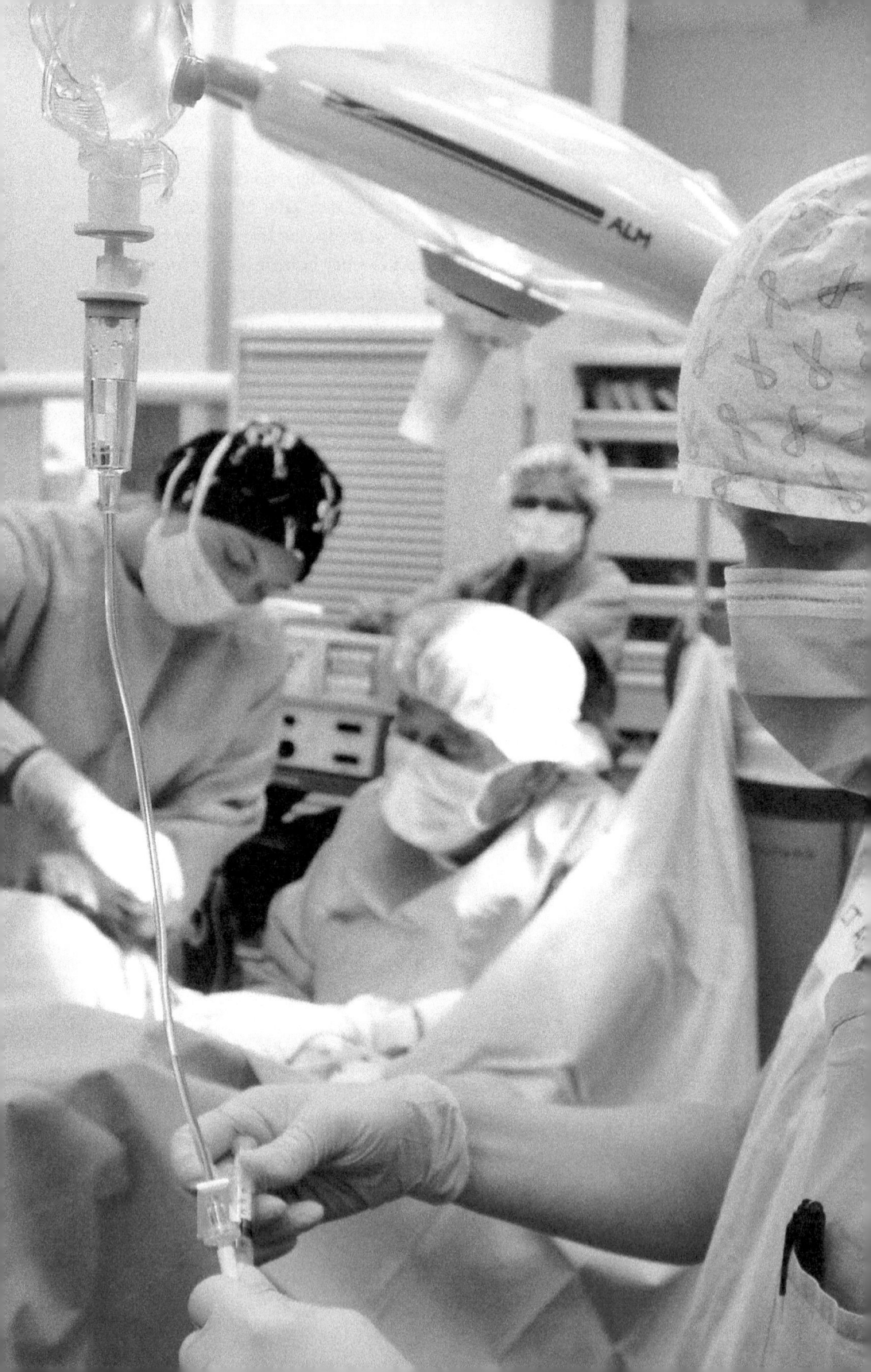

21

Surgery, Chemotherapy and Radiation Treatments

As a chemist trained to interpret data, it is incomprehensible to me that physicians can ignore the clear evidence that chemotherapy does much, much more harm than good.

Alan Nixon, Ph.D.
Past President, American Chemical Society

If you have been diagnosed with cancer by your oncologist it is most likely that conventional treatment will automatically recommend "the big 3 protocol" as in "Slash, Poison and Burn". That is, Surgery, Chemotherapy and Radiation treatment. They do not work effectively and their success rates are low, causing more harm than good to the human body. In most cases, the cancer is not totally removed from the body and patient survival rates are poor.

Surgery itself can often be one of the reasons why cancer cells spread! This could be due, for example, to a minute miscue or careless handling of tumour tissue by the surgeon. During surgery, in trying to remove the whole tumour, the surgeon in question can literally spill millions of cancer cells into a patient's bloodstream, and these can travel to other parts of the body.

In his book, "One Answer to Cancer", Dr. Donald Kelly wrote:

> Often, while performing a biopsy, the malignant tumour is cut across, and this tends to spread or accelerate cancer growth. Needle biopsies may accomplish the same tragic effects.[1]

The main concept of alternative treatments is that they visualize cancer as a multidimensional systemic total body disease. They attempt to find and correct the root cause of the disease, instead of just aiding the patient's symptoms of the disease. Dietary intervention treatments properly focus on cleansing the body and stimulating the natural immune system with specific diets, supplementations, detoxification and boosting oxygenation levels.

Dr. Philip E. Binzel, Jr., has documented the outcomes of treating his own cancer patients with using just natural supplementations and diet. He then compared his results to those shown by the American Cancer Society (ACS) for conventional treatments (surgery, chemotherapy and radiation). What he found out was astonishing.[2]

It is indeed a pity that studies on treating cancer with nutrition are infrequent – even though available data shows that the results of alternative natural treatments are overwhelmingly favorable: diet most certainly has a more powerful and more long-lasting beneficial effect than conventional treatments.

Patient Survival (5 years or more)

	Conventional Therapy	Nutritional Therapy
Primary Cancer	15%	87%*
Metastatic Cancer (spread)	0.10%	70%

* did not die of cancer and survived for more than 18 years

Primary cancer is defined as a detectable cancer that is confined to a single area (tissue), with perhaps a few adjacent lymph nodes involved.

Metastatic cancer is defined as a cancer that has spread to other parts of the body from the main site of growth, which can be located in multiple areas of the body.

The table above clearly shows that when it comes to curing cancer and survival rates, nutritional treatments are far superior and more effective than conventional treatments.[3]

History, we are told, often repeats itself. In an early study presented to the American Cancer Society (ACS) in the 1980's; concluded much the same as that study in France, over a century earlier.

In her book, "Forbidden Medicine", Ellen Hodgson Brown explains:

> One of the few studies (...) was conducted by Dr. Hardin Jones, professor of medical physics and physiology at the University of California, Berkeley. He told an ACS panel: (...) My studies have proven conclusively that untreated cancer victims actually live up to four times longer than treated individuals. For any specific type of cancer, people who refused treatment lived for an average of 12½ years. Those who accepted surgery or other kinds of treatment [chemotherapy, radiation, cobalt] lived an average of only three years. . . . I attribute this to the traumatic effect of surgery on the body's natural defence mechanism. The body has an amazing immune defence against every type of cancer.[4]

> **FACT BOX**
>
> *Remember, there are worse things than death: one of them is chemotherapy.*
> **Charles Huggins, M.D. (1966 Nobel Prize for his research on the relationship between hormones and prostate cancer)**
>
> *I will use treatment to help the sick. I will never use it to injure them or wrong them. I will not give poison to anyone.*
> **The Hippocratic Oath**

This defence mechanism is called a healthy immune system. Once the human immune system is immuno-compromised with conventional medicine, it can no longer fight its battle against cancer.

This is confirmed by Dr. Charlotte Gerson (The Gerson Institute):

"We get terminally ill patients untreated with conventional medicine, and they response much better, because the body has not yet been poisoned with the toxic chemicals used to treat cancer."

Radiation can induce the development DNA mutations and can also increase the risk of cancer growth. Chemotherapy is toxic, carcinogenic (causes cancer), destroys red blood cells, devastates the immune system and can also damage vital organs.

How Toxic is Chemotherapy and Radiation Treatment?

New cancers can be caused by chemotherapy or by secondary cancers. This "quaint" side-effect is often overlooked in the lists of side-effects in a drug's accompanying literature; though you can find this information quite easily from the National Cancer Institute.

Just to name a few, here are some of the toxic side-effects that can be caused by conventional cancer treatments:[5]

SURGERY, CHEMOTHERAPY AND RADIATION TREATMENTS

- Anaemia
- Bleeding problems
- Constipation
- Diarrhoea
- Fatigue (feeling weak)
- Hair loss
- Infection
- Memory and nerve disorders
- Mouth and throat changes
- Nausea and vomiting;
- Pain
- Sexual and fertility changes in both men and women;
- Swelling (edema)
- Urination changes

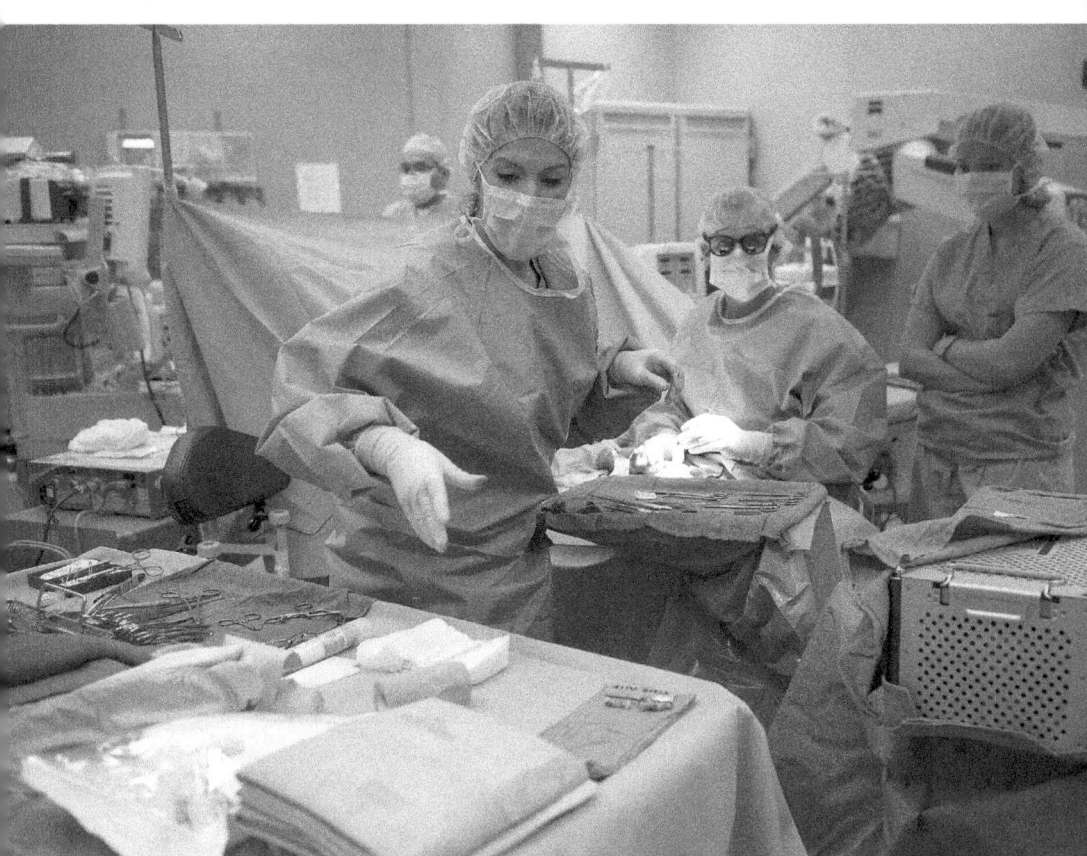

- Difficulty in swallowing
- Fever and chills
- Loss of appetite (Anorexia).

The following is reproduced from Mike Anderson's book, "Healing Cancer from inside out":

> The effectiveness of nutritional treatments scares conventional medicine and explains why they have fought so hard to deny funding for further study of diet. They are so anxious to protect their profitable toxic and ineffective treatments, that they have even rigged studies on dietary treatments and natural supplementation so they were doomed to fail.[6] Why are they so afraid? They are scared that people

> will abandon conventional treatments and flock to clinics which use dietary treatments. And they're right. If you have a choice between following a strict diet or chemotherapy, I can pretty much guess which choice you would make. If everyone knew just how ineffective conventional treatments really are, the choice would be a no-brain barrier. The issue here is that the cancer industry has to manipulate statistics and clinical trials[7] in order to make their treatments look better than what you would expect by taking a sugar pill (...)
>
> If the tumour is removed and the patient considered "cured", another tumour will most likely appear in another part of the body after a few years: because the treatment has done nothing to bolster the body's own defence systems and everything to destroy them. Within a year, the cancer might regain its power and start growing, long before the debilitated immune system is strong enough to combat the disease. This is maybe why 70 percent of women with breast cancer – who had the cancer show up in their lymph nodes – die from the illness. Their immune systems were wiped out as a result of the conventional treatment that could not contain the spreading cancer (...)

Here is what Hardin Jones said about the matter many decades ago and little has changed since then:[8]

> "My Studies have proven conclusively that untreated cancer victims live up to four times longer than treated individuals. If one has cancer and opts to do nothing at all, he will live longer and feel better than if he undergoes radiation, chemotherapy or surgery, other than when used in immediate life-threatening situations."

You might be now thinking, "but cancer treatments have now improved!" I am here to tell you they have not. And life-expectancy rates (long-term survival) following treatments have not budged one iota. The reason? Chemo and radiation both work by killing rapidly-dividing cells. Unfortunately, rapidly-dividing cancer cells only occur in a very small number of cancers (the very same reason chemo is somewhat effective in only a small number of cancers). In all other cases, cancer cells actually have slower division rates than the most active cells of the body. Thus, it is biologically impossible for Chemo to work without doing significant damage to other parts of the body. As a result, the strategy adopted in conventional medicine is to try to kill the cancer, before the treatment kills the patient. In the case of millions of people worldwide with cancer –that strategy did (and does) not work.

"Finding a cure for cancer is absolutely contraindicated by the profits of the cancer industry's chemotherapy, radiation and surgery cash trough." – Dr. John Diamond, M.D.

The future of breast cancer treatment should be primarily focusing on educating and empowering the general public about risk factors and risk awareness and giving out vital information, in relation, to the self-care strategies on how to improve dietary and lifestyle changes. These changes would most certainly safeguard public health in preventing breast cancer in the first place.

There is hope. That hope will be renowned whenever health-care systems truly accepts the facts: that diet-lifestyle interventions are solely the key determinants to prevent and reduce the risk of contracting cancer, and other diseases in all societies.

Notes for chapter 21

1. Kelly, W.D. (1974). *One Answer to Cancer*. Kelley Foundation. Revised Edition.
2. Binzel, P.E. (October 1994). *Alive and Well: One's doctor's Experience With Nutrition in the Treatment of Cancer patients*. p.107. Published by American Media. Moss, W.R. (1996). *Questioning Chemotherapy*. Published by Equinox Press. Anderson, M. (2009). Healing Cancer from inside out. Part 1: The Failure of Conventional Treatments: Results of Nutritional Therapy, pp 18-24. Published by www.RaveDiet.com
3. Ibid.
4. Hodgson Brown, E. (2008). *Forbidden Medicine: Is effective Non-toxic Cancer Treatment Being Suppressed?* Third Millennium Press, Second edition.
5. National Cancer Institute at the National Institutes of Health. Chemotherapy and Radiation side-effects. Retreived from http://www.cancer.gov/cancertopics/coping/physicaleffects/chemo-side-effects
6. See also Chapter 18, "Supplementations."
7. Clinical trials are supposed to be the "gold standard" of scientific study, and yet, it has been well documented that these studies almost always produce results that are beneficial to the organization providing the funding. The desires of the sponsors of the study, generally not using true scientific methods, almost always determine the outcomes of the study. This is accomplished through an elaborate system of fraudulent trial design, selective reporting, dismissing study subjects who do not produce the desired outcomes, statistical distortions and the application of "cancer pressure" to the researchers who carry out such studies (e.g. researchers who do not produce the desired results get fired or blackballed by the industry).
8. Transactions of the New York Academy of Medical Sciences, 1956, Vol.6.
9. Anderson, "Healing Cancer from inside out.", 2009.

22

Is There a Related Link Between Wearing a Bra and Breast Cancer?

The nature of the bra, the tightness, and the length of time worn, will all influence the degree of blockage of lymphatic drainage. Thus, wearing a bra might contribute to the development of breast cancer as a result of cutting off lymphatic drainage, so that toxic chemicals are trapped in the breast.

Dr. Michael Schachter
Director of the Schachter Center for Complementary Medicine, graduate of Columbia College of Physicians & Surgeons

No! I am not joking! In a book entitled "Dressed to Kill: The Link between Breast Cancer and Bras", Soma Grismailjer and Sydney Singer presented some startling statistics:[1]

- Women who rarely wear or never wear a bra had a 1 in 168 chance of developing breast cancer.
- Women wearing bras less than 12 hours a day had a 1 in 152 chance of developing breast cancer.
- Women wearing their bras more than 12 hours a day, but not in bed, have a 1 in 7 chance of developing breast cancer
- Women wearing a bra 24 hours a day had a 3 in 4 chance of developing the disease.

What could be the reason for this? According to Dr. David Williams:

> Wearing a bra at least 14 hours a day tends to increase the hormone prolactin, which decreases circulation in the breast tissue. Decreasing circulation can impede the body's natural removal of carcinogenic fluids that becomes trapped in the breast's lymph nodes. These glands consist of the largest mass of lymph nodes in the upper part of the body's lymphatic system.

Ty Bollinger explains theoretically what happens:[2]

> The restrictive nature of bras inhibits the lymphatic system (our draining system of vessels and nodes that helps flush away harmful waste from the body) from performing its task. The mammary glands have multiples of lymphatic vessels that run from the breast, through the auxiliary lymph nodes below the armpits, over to the collar bone and at length to the thoracic duct. Excess wastes and toxin build up can later enter the blood circulation to our liver, then the kidneys: additional – this cleansing and removing process can eliminate unwanted harmful waste via the bladder, and is eventually expelled with the urine, keeping our internal environment clean.
>
> However, if something impedes the cleaning process, an imbalance occurs. If there is a build-up of harmful chemicals (carcinogenic agents) and other oestrogen by-products, this can create destructive molecules

IS THERE A RELATED LINK BETWEEN WEARING A BRA AND BREAST CANCER?

called free radicals, which can accumulate, leading to cellular damage and the increased risk of developing breast cancer.[3]

If you cannot discontinue wearing a bra, try to consider wearing one only whenever needed. In addition, women should try to wear a more appropriate-sized bra that allows some breast motion, without cutting tightly under and along the outer edges of the breasts where the

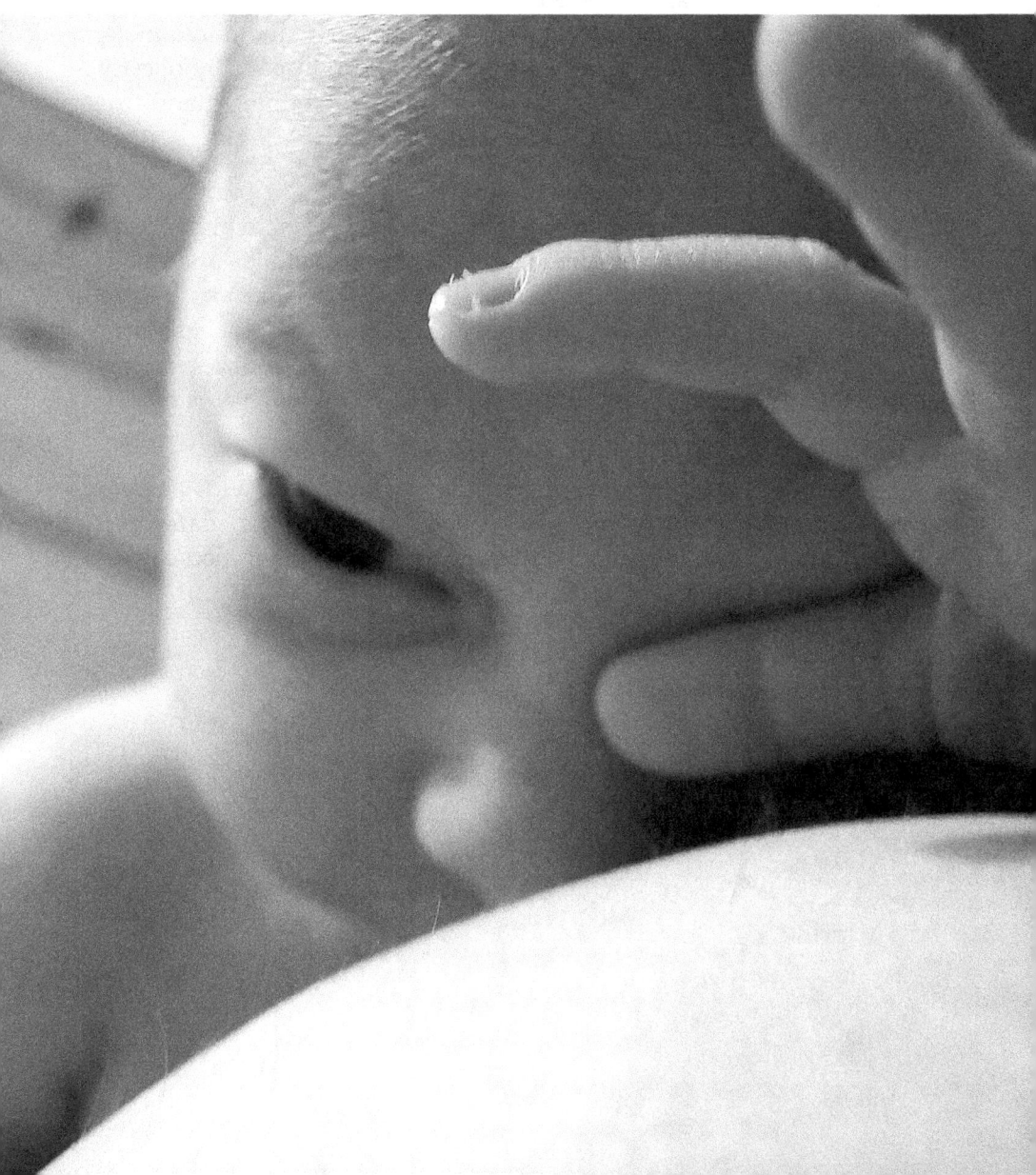

milk ducts are located. Body size, weight and body fat have also been associated with the increased risk of breast cancer.[4]

In their observations (Singer and Grismaijer) may have not included other known risk factors for breast cancer: diet, exercise, lifestyle, body weight, hormone levels, start of menstruation and breast feeding. All in all, the combinations of all these risk factors may have tipped the iceberg and brought upon the disease, and *not* by just-wearing a bra![5]

Bra cup size and hardness were also studied, as possible risk factors for breast cancer.[6] Although, the association was found only among postmenopausal women and was accounted for, in part, by obesity. Data suggest that bra cup size (and conceivably mammary gland size) may be a risk factor for breast cancer.

However, the evidence does conclude that wearing a bra decreases circulation in the breast tissue. Again, decreased circulation can impede the body's natural removal of carcinogenic fluids and toxins that become trapped in the breast's lymph nodes that are, indeed, hazardous to human health. Simply put, we cannot ignore these facts. In other words, not wearing a bra will not stop cancer from developing, although it can increase the risk of the disease; although, it is not the main cause of cancer.

The cure for cancer is preventing it in the first place with obvious changes: that is, try not to expose ourselves to dangerous environmental and man-made toxins, making healthier food choices and changing one's lifestyle.

Notes for chapter 22

1. Singer, S.R. & Grismaijer, S. (2005). *Dressed To Kill: The link between breast cancer and bras*. Published by Iscd Pr, 1st ISCD Press ed edition.
2. Bollinger, Ty. (2011). *Cancer: Step outside the box*. Chapter 8: Common Cancers and Cachexia. Wearing a Bra, pp.192-193. Published by Infinity 510^2 Partners.
3. Ibid.
4. Ibid.
5. Bollinger, "Cancer step outside the box", 2011.
6. Hsieh, C.C *et al*. (1991). Breast size, handedness and breast cancer risk. *European Journal of Cancer*, **27**(2), pp. 131-135. DOI: 10.1016/0277-5379(91)90469-T

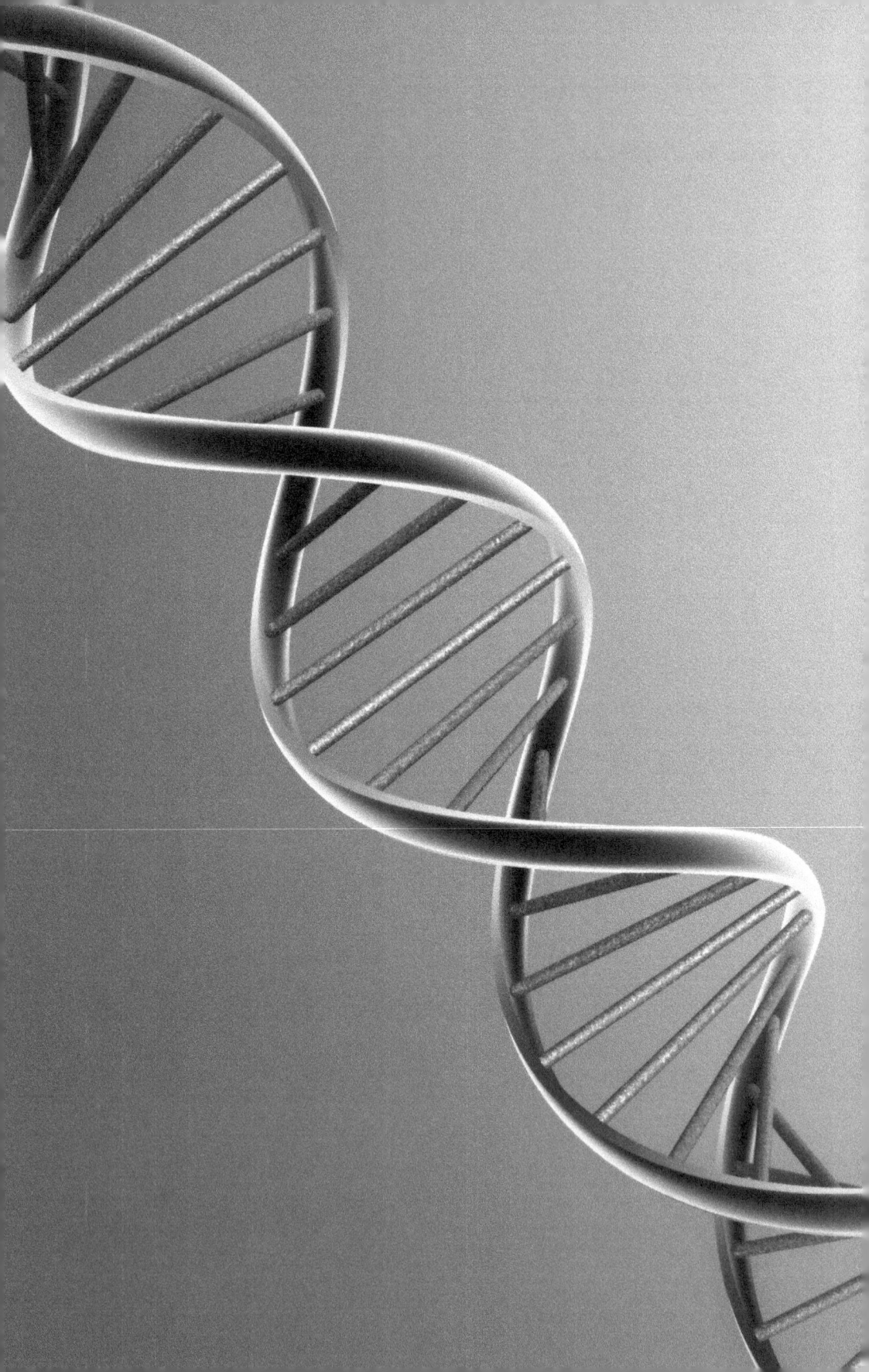

23

Genes do *not* Determine Disease on their Own

Your genetics load the gun.
Your lifestyle pulls the trigger.

Mehmet Oz
Turkish-American cardiothoracic surgeon, author, and television personality

Genes: Health or Disease

The origin of every single disease is genetic. Genes are the code to everything in our bodies, good or bad. Without genes, there would be no cancer. Without genes, there would be no diabetes or heart disease. And without genes, there would be no life.

Much of the focus on genes, however, misses a simple but critical point: not all genes are fully expressed all the time. If they are not activated, or expressed, they remain biochemically dormant. Dormant genes do NOT have any effect on our health. This is a known fact to most scientists and many lay people, but the full significance of this idea is seldom understood. What causes some genes to remain dormant and others to express themselves? The answer: environmental pollutants, especially diet.[1]

Genetic links to cancer only present a very small fraction of most cancers (2% to 5%) including breast cancer. Yet, many people wrongly blame the development of cancer on their genes. It is important to note that:

- The risks for people whose genes are linked (inherited) to cancer can be drastically reduced just by making simple dietary and lifestyle changes, like those contained in this book.
- People have been misinformed by articles in the press, written by doctors, health care system officials and cancer organizations.
- Because of this widespread misinformation, people start to look at other factors (such as genes and environmental pollutants) as having contributed to or caused their cancer.

All humans are born with a matrix of genes, which can make us more or less susceptible and prone to disease. It is true that genes inherited from our parents and from our ancestry tree can be genetically passed down to us. However, mutated genes may or may not become active unless the right conditions trigger them off. Yes! Genes load the gun, but lifestyle pulls the trigger.[2]

A Healthy "Genie" Advice by Dr. John A. McDougall

There is a genetic predisposition to all diseases, it is true. But evidently, most people can still avoid the development of disease, if they do not stress their genes with unhealthy habits.

In a documentary entitled "Healing cancer from inside out" Dr. John A. McDougall gives this advice:

> You can present some hereditary strengths and weaknesses in life and so then you do harmful things to your body: smoke, drink alcohol, and eat unhealthy foods. These genetic strengths or weaknesses may express themselves. But, if you at no time smoked, drank alcohol, ate unwell, or inhaled harmful toxins you would have never known that you had a tendency for colon cancer, diabetes, lung cancer, liver disease, breast cancer or prostate cancer, etc. You would not have found out these transmissible weaknesses because these health problems do not occur in populations of people who eat a diet based on starch, vegetables and fruits. In scientific literature you do not see people inflicted with the above ailments; nobody has them. In China, there are hundreds of millions of people, and not anyone has these diseases, or they are very rare indeed.

One of the more exciting benefits of good nutrition is the prevention of diseases that are thought to be due to inherited predisposition. We now know that we can largely avoid these diseases, even though we may have inherited or harbored the gene (or genes) that is (or are) responsible for the disease.

However, billions of dollars go into funding genetic research which continues to spiral upwards in the belief that genetically inherited or acquired genes account for the occurrence of specific diseases. A hope that someday and somehow genetic research on the human genome will be able to find a miracle cure in "turning off" these nasty genes.

Drug company public relations programs now depict a future where each one of us will have a personal ID card cataloguing all our good and bad genes. Using this card, we will be expected to go to the doctor who will prescribe a single pill to suppress our harmful genes. I strongly believe that such "miracles" will never be realized. If indeed such experiments are tried, I am sure they may create serious and undesired consequences to human health. Why? Unfortunately, futuristic pipe dreams obscure the affordable, efficacious health solutions that already exist: solutions based on nutrition.³

A healthy diet and lifestyle will keep bad inherited genes inactive, such that nasty genes will not be triggered or awakened. Most diseases can be attributed to a Western diet, whereby people eat more processed, more refined, and larger portions of animal products and not enough powerful protective nutrients, that are mainly found in the abundance of plant foods.

Mike Anderson states that "What a strong family history suggests is a history of bad diet, and lifestyle can awaken these genes".

And T. C. Campbell (of Cornwell University) confirms:

> All of us have bad genes. No human being on this planet has perfect genes: Genes, which we all wish we never inherited or acquired later. Either we were born with them, or they were corrupted during our lifetime, by exposure to certain agents. So, we all have genes, that can cause us health problems. And these damaged genes *will* without any doubt, sometime or another, during our lifetime, makes us more susceptible to disease – if, of course, we continue to fertilize them. Nutrients control the expression of genes, but if we do not fertilize these genes with the wrong kinds of nutrient's and lifestyle, our health will prevail in keeping these genes under control or dormant.

If you have genetically inherited from your parents or grandparents the genes for breast cancer (BRCA-1 or BRCA-2) – these genes, are more likely to be passed down to your sons or daughters. Remember that what parents eat and consider healthy eating, the more likely that their children will follow in their footsteps, and follow the same diet – good or bad. Basically, the risk of getting breast cancer will increase, especially if you have inherited the genes for breast cancer. In most cases, disease in family generations follows disease – if, of-course, that family's diet and lifestyle do not improve or change.⁴

GENES DO *NOT* DETERMINE DISEASE ON THEIR OWN

Among the genes that influence breast cancer risk, BRCA-1 and BRCA-2 have received the most attention since their discovery in 1994.[5] These genes, when mutated, confer a higher risk for both breast and ovarian cancers.[6] These mutated genes may be passed on from generation to generation, that is, they have inherited genes.

In the excitement over these discoveries, however, other information has been ignored:

1. Only 0.2% (1 in 500) of individuals in the general population carry the mutated form of these genes.[7] Because of the rarity of these genetic aberrations, only a small percentage of breast cancer cases in the general population can be attributed to mutated BRCA-1 or BRCA-2 genes.[8]
2. These genes are not the only genes that participate in the development of this disease,[9] and many more will surely be discovered.
3. The mere presence of BRCA-1, BRCA-2, or any other breast cancer gene, does not guarantee disease occurrence. Environmental and dietary factors play a central and critical role in determining whether or not these genes are expressed.[10]

It is worth mentioning that most of the women who inherit the genes for breast cancer do not develop the disease. 90-95 percent of women who are diagnosed with breast cancer do not have the BRCA-1 or BRCA-2 inherited genes, and only about 5% of breast cancers run in families.[11] Women with these genes certainly face higher risks for breast cancer.[12] But even among these high-risk women who have the gene, there still have a good reason to believe that more attention to diet will be more likely to pay handsome rewards. Evidently, about half of the women who do carry these "rare potent-genes", do NOT develop breast cancer.[13]

Again, as already mentioned, whenever people migrate from areas of the world where disease incidence is low to areas of the world where incidence is high, they quickly adopt the high incidence rates as they change their diet and lifestyle.[14] This shows that even though individuals may have the inherited gene(s), the disease will occur only in response to certain dietary and/or environmental circumstances. The same goes for cancer, heart disease and Type 2 diabetes. People acquire the risk of the population to which they move, especially if they move before their adolescent years.[15] So, this is evidence that disease is more correlated with environmental factors and lifestyle changes, than it is with genes.[16]

This proves that in most cases the inherited genes are not to be blamed, as other factors are involved. A healthy diet and lifestyle are preventive factors against inherited gene expression in chronic-related diseases, including breast cancer."

We need to remind ourselves that these genes need to be "expressed" in order for them to participate in disease formation, and nutrition can affect this.

I think the following quote from Mike Anderson' book, "Healing Cancer from inside out" is very pertinent: "Since bad genes awaken due to changes in the body's environment, you can put them back to sleep with changes in diet and lifestyle".

So the genes we inherit and changes in diet can affect DNA expression at a genetic level. In other words, or rather in G. Michael's words:

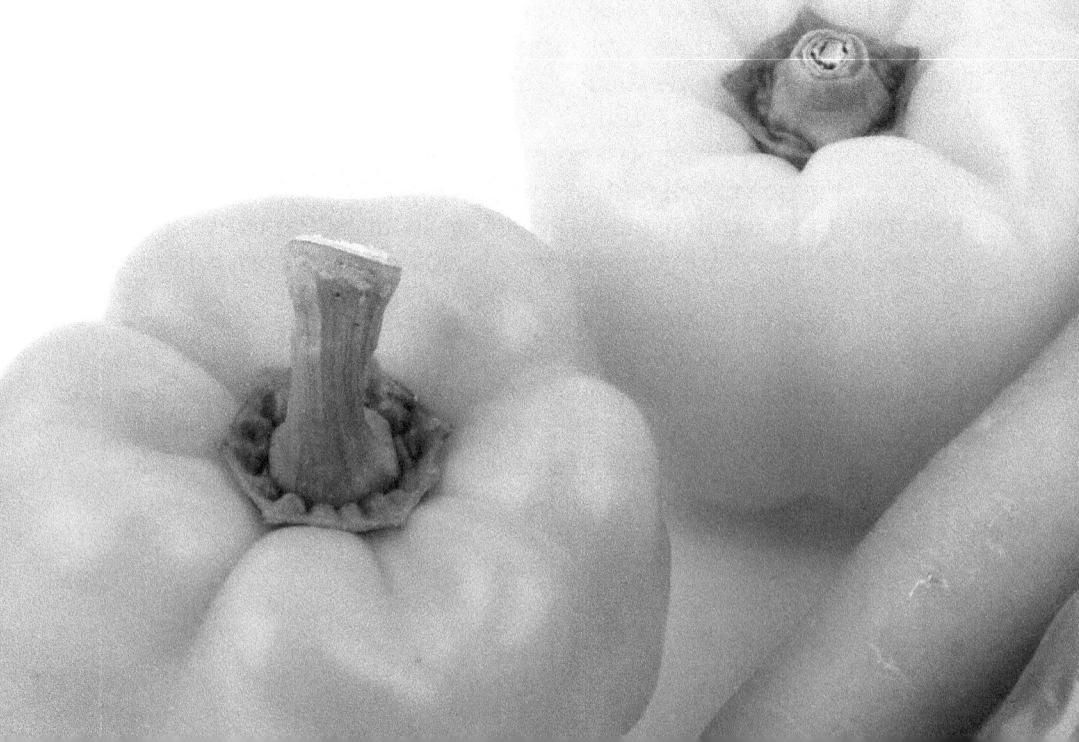

"No matter what bad genetic cards we've been dealt, we can reshuffle the deck with diet."[17]

This means that our genetically inherited genes are *not* the main reason why many women and a small percentage of men develop breast cancer: Diet and environmental factors are high-risk factors of gene expression and disease development.

For example:

If you have two people living in the same environment and you feed them exactly the same meaty food every day for their entire lives, I would not be surprised if one died of cancer at the age of eighty and the other person died of a heart attack at age fifty-five. What would explain these differences in disease occurrence and disease development? Well, genes give us our predispositions. We all have different disease risks due to our different genes. So, while we will never know exactly which risks we are predisposed to, we do know how to control these risks. Regardless of our genes, we can all optimize our chances of expressing the right genes by providing our bodies with the best possible environments – that is, the best possible nutrition and other healthy lifestyle interventions. Even though the two different persons in the example above succumbed to different diseases at different ages, it is entirely possible that both could have lived many

more years with a higher quality of life, if they would have practiced optimal nutrition and bettering their lifestyles.

However, there are exceptions to these rules. Some people are just entirely more health resilient: having healthier genes, a stronger immune system and can also cope with life's worries and stresses much better than most people. These persons are in a small majority that are exceptions to the rule of disease gene expression. So, for these few exceptional people, there is a greater tendency that they *will* live a healthy and longer life.

But, for the majority of the general population – whenever they choose to eat a bad diet and enroll in a bad lifestyle, they certainly do indeed suffer the consequences of developing diseases sooner and faster, than those few who are more health resilient.

We all must try to eat foods that will protect us from environmental toxins. Even though we do not know which genes we have for a specific disease; we all should try to protect all our genes from alternations or mutations, that may lead to gene expression and disease development.. This can be accomplished by eating foods which will automatically protect our genes, and choosing to change the way we live our lives.

The traditional model of how fruits and vegetables protect us from cancer, is that their antioxidants prevent the build-up of free radicals, also known as reactive oxygen species (ROS), which would otherwise go on to damage our cellular DNA, cell membranes, mitcohrondia, etc……. which could lead to the transformation of healthy cells into uncontrolled, damaged, diseased, or dying cells.[18]

Gene regulation can be controlled by plant foods. A study performed on berries and other plant-based foods, concluded that:[19]

"The observed changes in blood cell gene expression profiles suggest that the beneficial effects, of a plant-based diet on human health may be mediated through optimization of disease processes." So the phytonutrients found in the abundance of plant foods may actually modulate gene expression and can increase our cellular defenses, such that even if there is some damage to our DNA our cells may recover, instead of being irreparably lost. This may indeed reduce our risk of bad genes becoming expressed and disease developing.[20]

Genes do indeed control our lives, but a healthy plant-based diet and other lifestyle interventions, can most certainly hinder inherited or acquired damaged genes from turning against us.

Notes for chapter 23

1. Campbell, T.C. & Campbell, T.M. (2006). *The China Study.Part III: The Good Nutrition Guide: Chapter 11: Eating Right: Eight Principles of Food and Health.* pp.233. Published by BenBella Books, Inc.

2. Esselstyn, C.B. (2010). Is the Present Therapy for Coronary Heart Disease the Radical Mastectomy of the Twenty-First Century? *American Journal of Cardiology,* pp.902-904. doi:10.1016/j.amjcard.2010.05.016

3. Campbell, T.C. & Campbell, T.M., op. cit. pp.161-164.

4. Ibid.

5. Liu, Q. et al.(1994). BRCA1 mutations in primary breast and ovarian carcinomas. *Science,* **266**, pp.120-122; Wooster, R., et al. (1995). Identification of the breast cancer susceptibility gene BRCA2. *Nature,* **378**, pp.789-792.

6. Ford, D. et al. (1994). Risks of cancer in BRCA1 mutation carriers. *The Lancet,* **343**, pp.692-695.

7. Ibid.

8. Newman, B. et al. (1998). "Frequency of breast cancer attributable to BRCA1 in a population-based series of American women. *JAMA,* **279**, pp.915-921. Petro. J., et al. (1999). Prevalence of BRCA1 and BRCA2 gene mutations in patients with early-onset breast cancer. *Journal of the National Cancer Institute,* **91**, pp. 943-949.

9. Ibid.

10. Campbell, T.C. ibid.

11. Shauna, N et al. (March 2009). CpG Island Tumour Suppressor Promote Methylation in Non-BRCA-Associated Early Mammary Carcinogenesis. *Cancer Epidemiology Biomarkers Prev,* **18**(3), pp.901-914. doi: 10.1158/1055-9965.EPI-08-0875.

12. Antoniou, A et al. (2003). Average risks of breast cancer and ovarian cancer association with BRCA1 or BRCA2 mutations detected in case series unselected for family history: a combined analysis of 22 studies. *American Journal of Human Genetics,* **72**, pp.1117-1130.

13. Ibid.

14. Stephenson, C et al. (1992). Evidence for an environmental effect in the aetiology of insulin-dependent diabetes in a transmigratory population. *British Medical Journal,* **304**, pp.1020-1022. Burden, A.C., Samanta, A., & Chaunduri, K.H. (1990). The prevalence and incidence of insulin-dependent diabetes in white and Indian children in Leicester City (UK). *International Journal of Diabetes Dev Countries,* **10**, pp.8-10. Ellot, R., & Ong, T.J.

(2002). Nutritional genomics. *British Journal of Medical*, **324**, pp.1438-1442. Campbell, T.C. & Campbell, T.M., op. cit., p.191.

15 Acheson, E.D *et al*. (1966). Multiple sclerosis: a reappraisal. *JAMA*, **194**(8), pp.939. doi:10.1001/jama.1965.03090210103054. Alter M, *et al*. (1966). Risk of multiple sclerosis related to age at immigration to Israel. *Arch Neurol*, **15**(3), pp.234-237. doi:10.1001/archneur.1966.00470150012002.

16 Kurtzke, J.F *et al*. (1979). Epidemiology of multiple sclerosis in U.S veterans: 1. Race, sex, and geographic distribution. *Neurology*, **29**, pp.1228-1235. Campbell, T.C. & Campbell, T.M., op. cit., p.198.

17 Greger, M. (2014). *Can Eating Soy Prevent Breast Cancer?* Retreived from http://nutritionfacts.org/2014/09/18/can-eating-soy-prevent-breast-cancer/

18 Greger, M. (July 19th 2012). Plant-based diets and cellular stress defences. Volume 9. Retreived from http://nutritionfacts.org/video/plant-based-diets-and-cellular-stress-defenses/

19 Bohn *et al*. (2010). Blood cell gene expression associated with cellular stress defense is modulated by antioxidant-rich food in a ramdonized controlled clinical trial of male smokers. *BMC Medicine*, **8**(54). doi: 10.1186/1741-7015-8-54.

20 Greger, "Plant-based diets", 2012.

24

Cow's Milk is the Perfect Food for Baby Calves. It is *not* Healthy for Humans

The human body has no more need for cows' milk than it does for dogs' milk, horses' milk, or giraffes' milk.

Michael A. Klaper, M.D.
American physician, author, and vegan

The Moment has Come to Fully Elucidate why "Cow's Milk" is a Detrimental Food Choice for Mankind

Most of us have been raised with the idea that cow's milk is nutritious and a healthy food. But cancer researchers are now showing us the down-side of dairy products that might surprise us. But what, exactly, is in cow's milk?

If you take a typical glass of milk to a clinical laboratory and test its constituency, the first thing you will discover is that around 49% of the calories are nothing but fat. Most of this fat is in fact saturated fat, or bad fat, including small amounts of cholesterol. One cup of whole milk contains 33 milligrams of cholesterol, which indeed has been proven to increase your cholesterol levels.

What about skinned milk? The problem with skinned milk is that when you take away most of the fat, around 55% of skinned milk actually consists of the disaccharide sugar lactose, (derived from galactose and glucose), and can be problematic for many people. A single cup of low-fat milk also contains approximately 10 mgs of cholesterol. The other components of milk mainly consist of bovine hormones and high amounts of the protein casein. We will discover later, milk is a nutritional food source only intended for baby calf's to drink. If humans drink cow's milk there will be many health implications, including the risk of developing cancer.

The Trouble with "Whey" Protein

Some health-conscious individuals and body builders take whey protein to help them build muscles. What is whey protein, and is whey a health concern?

After fat and casein are removed from cows' milk, dairy processors are left with whey protein. Whey is composed of bovine blood proteins. These include serum albumen, dead white blood cells, and hormonal residues like oestrogen and progesterone, etc. The way the human body reacts to a foreign protein is to bring in its immune system: which may attack the foreign entity or the antigen-like invader. In the case of some individuals (with a predisposition), the immune system starts to attack itself. Cow and other animal proteins and tissues (hormones, cartridge, bone, tendons, ligaments, etc), are very similar to our own cells and tissues. It now has been hypothesized that autoantibodies may be produced by the human immune system, and these autoantibodies my indeed react or attack its own cells and tissues. This is known as an "auto-immune response". This may be how autoimmune diseases come about: molecular mimicry and/or loss of immunological tolerance, the ability for an individual to discriminate between self and non-self.

Robert Cohen, author and executive director of "MILK A-Z" clearly describes what happens:

> In the case of diabetes and Multiple Sclerosis (MS), the body's response to whey proteins is to attack the outer membrane protecting nerve cells, or the myelin sheath. It has long been established that early exposure to bovine proteins is a trigger for insulin-dependent diabetes mellitus (IDDM). Researchers have made that same milk-consumption connection to MS. The July 30, 1992 issue of *The New England Journal of Medicine* first reported the diabetes autoimmune response milk connection. Patients with insulin-dependent diabetes mellitus produce antibodies to cow milk proteins that participate in the development of islet dysfunction (...) Taken as a whole, our findings suggest that an active response in patients with IDDM (to the bovine protein) is a feature of the auto-immune response.[1]

On 14 December, 1996, *The Lancet* revealed that:

> Cow's milk proteins are unique in one respect: in industrialized countries they are the first foreign proteins entering the infant gut, since most formulations for babies are cow milk-based. The first pilot stage of our IDD prevention study found that oral exposure to dairy milk proteins in infancy resulted in both cellular and immune response (...) This suggests the possible importance of the gut immune system to the pathogenesis of IDD.

The Multiple Sclerosis/Milk Connection

The April 1st 2001 issue of the *Journal of Immunology* contained a study linking MS to milk consumption.

Michael Dosch, M.D. and his team of researchers determined that MS and type I (juvenile) diabetes mellitus are far more closely linked than was previously thought. Dosch attributes exposure to cow milk protein as a risk factor in the development of both diseases for people who are genetically susceptible. According to Dosch:

> We found that immunologically, type I diabetes and MS are almost the same – in a test tube you can barely tell the two diseases apart. We establish that the autoimmunity was not specific to the organ system affected by the disease. Previously it was thought that in MS autoimmunity would develop in the central nervous system, and in diabetes it would only be found in the pancreas. We found that both tissues are targeted in each disease.

Women are targeted by dairy industry scare tactics that offer misinformation regarding osteoporosis. Two-thirds of MS victims are adult females.

As milk and cheese consumption increase along population lines, so too does an epidemic number of MS cases.[2] The numbers add up. The clues all add up. Science supports epidemiological studies.[3]

The Problem with the Milk Sugar "Lactose"

Lactose intolerance is common among many populations: affecting approximately 95% of Asian Americans, 74% of Native Americans, 70% of African Americans, 53% of Mexican Americans and 15% of Caucasians.[4] Symptoms, which include gastrointestinal distress, diarrhea and flatulence occur, because these individuals do not have the enzymes that digest the milk-sugar lactose. Additionally, along with unwanted symptoms, milk-drinkers are also putting themselves at risk of developing other chronic diseases.

When women consumed more milk products, especially yoghurt, they had been implicated in ovarian cancer.[5]

Dr. D.W. Cramer suggested that the culprit might be a normal breakdown product of the milk sugar, lactose. Lactose is broken down in the body to another sugar called galactose. In turn, galactose is broken down further by enzymes in the body.

According to Dr. Cramer, "whenever dairy product consumption exceeds the enzymes' capacity to break down galactose, there is a build-up of galactose in the blood, which can damage a woman's ovaries".

Some people have particularly low levels of the enzymes needed to break down galactose, and when they consume dairy products on a regular basis, their risk of ovarian cancer could be triple that of other non-dairy-consuming women.

Cramer noted that the primary problem was the milk sugar (lactose) and not the milk fat, thus the trouble is not solved by using low-fat dairy products. Actually, cottage cheese (ricotta) and yoghurt seem to be of most concern, because the bacteria used in their production increase the production of galactose from lactose.[6]

Other studies have also linked dairy products to ovarian cancer.[7]

Proteins and Hormones in Cow's Milk

Another problem is that scientists have theorized that the proteins and other hormonal constituents in milk can trigger allergies,[8] diabetes,[9] osteoporosis,[10] cardiovascular disease,[11] autoimmune diseases,[12] including cancer.[13]

Countries that consume more milk have been shown to have more breast and prostate cancer (United States, Europe, Australia and other westernized countries) compared to countries where milk is not a big part of their daily-diet: China, Thailand and Japan. But, perspectively (subjective evaluation of relative significance: a point of view), countries and other regions around the world that have been copying the westernized dietary pattern, these diseases have increased.[14]

Consuming More Dairy Products May Lead to a Higher Risk of Osteoporosis?

Dairy products may actually increase our risk of osteoporosis. The many studies linking intake of animal protein to bone loss, whilst also showing a worse calcium balance with increased dairy intake, show how unfounded commercials are when they promote dairy products (daily calcium intake) as a positive approach towards healthy bones and teeth.

Whenever persons get their protein from plant sources, instead of animal products, bones tend to breathe a sigh of relief. Why? Animal products are high in what are called *sulphur-containing amino acids*.[15] These acidic amino acids (Methionine and Cysteine), found in higher concentrations in animal proteins, tend to leach calcium from the

bones, and that calcium passes through the kidneys and out in the urine.[16]

The human body metabolizes sulphur-containing amino acids, methionine included, into sulphuric acid, which is one of the most potent acids found in nature. These potent dietary acids dissolve the bones and may cause the kidneys to produce calcium-based stones.[17]

While still containing all the essential amino acids the body needs for repairing and building body tissues, plant protein sources are far lower in sulphur-containing amino acids and help to protect human bones. Overall, plant foods are more alkaline-forming foods.

Dairy products, especially cheeses are high in protein and acid. The health consequence is that some of these amino acids (sulphur-containing amino acids) found in the protein casein can be digested and further broken down to cause an acidic environment in the body that may cause osteoporosis, which ultimately leads to calcium loss from the bones.

The high protein content of milk may also damage the kidneys (kidney overload) which may contribute to the development of kidney stones and kidney disease. A low-protein diet is recommended for anyone who has bone and kidney problems.

Milk and Countries with the Highest Rates of Osteoporosis

According to Dr. John McDougall:

- The countries with the *highest consumption of dairy products* are: Finland, Sweden, United States of America and England.
- The countries with the *highest rates of osteoporosis* are: Finland, Sweden, United States of America and England.[18]

Dairy products are also high in cholesterol, which can promote the development of atherosclerosis, strokes and heart attacks. Dairy products are also very low in iron, contain no Vitamin C and dietary fibre: that maybe the cause of anaemia, scurvy, and constipation, causing inflammation of the bowels leading to anus fissures and pain. Other bowel conditions that have been studied and linked to the consumption of dairy products have been noted: Colitis, Ulcerative colitis, Crohn's Disease and the formation of bloody stools.

Dairy products are deficient in essential fatty acids (omega-3 and omeg-6 fatty acids) and contain harmful saturated fats, which may lead to the development of a weakened nervous system and a contributing factor to the risk of developing MS, Parkinson's disease, autism, headaches, acne, fatigue and heart-related conditions.

Rhinitis (inflammation of the nose), gastroesophageal reflux (backflow of stomach acid), allergies, asthma and/or eczema can be traced to the unnatural habit of drinking cow's milk.

Bones and joint problems have been associated with consuming dairy products – non-specific arthritis, rheumatoid and juvenile arthritis, including Lupus erythematosus.

So, the problem with retaining calcium from cow's milk is that you also get some or all of the baggage found in milk (saturated fat, cholesterol, lactose, casein and casomorphin, bovine hormones, antibiotics and pus cells (white blood cells) that are contributing factors to ill health. Overall cow's milk is *not* beneficial to human health.

We must remember that what the mother eats or drinks will ultimately affect her baby's health.

Thankfully, some parents are waking up to the truth about dairy products and seeking healthier milk alternatives for their families. Just as no kind parent would allow his or her small child to play on busy highways, drink alcohol or smoke cigarettes. Dairy products should be approached with extreme caution; given the amount of information readily available about their dangers.

As parents, we have the responsibility to research and educate ourselves about the food choices we offer our children. They count on us wholeheartedly for looking out after their best interests. Helping children avoid dairy products is one of the most loving, caring, and responsible actions parents can take for their children, for themselves, for society and of course to the health of cows and for baby calfs.

The Calcium Myth

"An important fact to remember is that all natural diets, including purely vegetarian diets without a hint of dairy products, contain adequate amounts of calcium that are above the threshold for meeting your nutritional needs. In fact, calcium deficiency caused by an insufficient amount of calcium in the diet is not known to occur in humans." – John McDougall, M.D.

There is a healthier substitute for cow's milk in obtaining calcium in one's diet. Calcium obtained from green-leafy vegetables, nuts and seeds can be better absorbed and bio-available for humans. There are many healthy plant-based foods you can choose from that contain adequate amounts of calcium: green-leafy vegetables like spinach, broccoli, Brussels sprouts, kale, turnips, and also legumes, almonds, sesame seeds and other nuts and seeds which do not contain the extra harmful-baggage that can cause health problems in humans.

However, it is important to remember to expose yourself to daily sunlight or add Vitamin D supplements to your daily diet, in order to increase calcium absorption/bioavailability.

Exercise More than Diet Strengthens Bones

Osteoporosis is *not* genetic and you don't have to suffer bone loss if you exercise and improve your eating habits. When people say it runs in the family, it simply means the family had led a sedentary lifestyle over generations. Bones are just like muscles. The single best thing you can do to prevent (and reverse) osteoporosis is to start exercising your bones, *because lack of physical activity is the primary cause of osteoporosis.*[19]

It's never been proven that a plant-based diet alone can reverse osteoporosis. However, it has been proven that exercise reverses osteoporosis. There are hundreds of studies which show far more dramatic differences in bone density between those who get regular exercise and those who do not.[20]

Acidic diets will cause bone loss, but not as rapidly as a lack of exercise. People who argue that acidic diets are the primary causes of bone loss have a big flaw in their argument: because there are still many meat-loving omnivores that have strong and healthy bones. This is because their activity levels strengthen their bones, despite a bad diet. Also, obese women and men never get osteoporosis because they get lots of weight-bearing exercise just carrying their bodies around: regardless of their diet and calcium intake. Lighter persons have increased chances of osteoporosis simply because their bodies don't have enough weight –unless they run or jog to add stress to their bones. In other words, lighter men and women need to engage in weight-bearing activity in order to strengthen their bones.[21]

The Correlation Between Dairy Products, Hip Fractures and Osteoporosis

Milk is one the best sources of calcium, but because of the lactose and animal proteins that it contains, animal products can cause an acidic state in one's body, which can lead to a higher rate of calcium being released from our "muscles or bones" and excreted in the urine. Dairy products are poor choices for calcium mainly because they *do* contain excessively high amounts of protein. Cow's milk has over three times more protein compared to human-breast milk. It was, after all, designed for a calf that will weigh over 300 pounds within a year of its birth. Whenever you consume dairy products, the calcium simply passes through your body and ends up in the toilet – the human body becomes more acidic because of bovine proteins. Low-fat dairy products are higher in protein, so you may lower your fat intake, but you will raise your blood acid level and this may contribute to excess bone loss.

Skim milk, for example, contains almost twice the amount of protein as whole milk. And because of the high sodium content in dairy products, they provide a double-boost to your blood acid levels.

In short, consuming dairy products turns out to be the worst possible way to build strong bones. You don't need the extra fat, the sodium, the

COW'S MILK IS THE PERFECT FOOD FOR BABY CALVES. IT IS *NOT* HEALTHY FOR HUMANS

acidic proteins, the cholesterol and you certainly don't need the excess calcium, which can result in painful kidney stones.[22]

The World Health Organization has yet to document a single case of calcium deficiency anywhere in the world.[23] In fact, there has not been a single recorded case of calcium deficiency of a dietary origin in the history of the entire planet.[24] Yet, the dairy industry tells Americans, they have a calcium deficiency.

Hip fractures and osteoporosis are more prevalent in populations in which dairy products are commonly consumed and where calcium intakes are commonly high.[25]

American woman drink thirty to thirty-two times as much cow's milk as the New Guineans, yet suffer forty-seven times as many broken hips. A multi-country analysis of hip-fracture incidence and dairy-product consumption found that milk consumption has a high statistical association with higher rates of hip fractures.

But, does this suggest that drinking cow's milk causes osteoporosis? Yes! However, it also could be due to the protein sources of milk and that increases the acidic load within the body, which in-turn actually removes more calcium from the skeleton, leading to bone weakness and osteoporosis. Let's get back to this answer later-on in this chapter, and try to answer this question with more evidence-based-studies.

Another major finding based on the Nurses' Health study, which included 121,701 women ages between 30 -35 at enrolment in 1976. The study estimated that the data does not support the hypothesis that the consumption of milk protects against hip or forearm fractures.[26]

In fact, those women who drank three or more servings of milk a day had a slightly higher rate of fractures than women who drank little or no milk. This does not mean that dairy products are not protecting us from osteoporosis as we have been indoctrinated to believe since childhood, but, on the contrary – studies show fruits and vegetables are essentially more protective against osteoporosis. Could it be that fruits and vegetables have a lower-acidic load, but also contain all the calcium we need in our diets, that overall aids against bone loss?

Osteoporosis has a complex aetiology that involves other factors, such as dietary acid-alkaline balance, trace minerals, phytochemicals in plants, exercise, adequate exposure to sunlight, and more.

Dr. Campbell, head of nutritional research for the *China Study Project*, also reported,

"Ironically, osteoporosis tends to occur in countries where calcium intakes are of the highest and most of it comes from protein-rich dairy products. The Chinese data gathered up from many years during the China Study indicates that people need less calcium that they think and also can get adequate amounts from many healthy sources of plant foods: kale, cabbage, broccoli, Brussels sprouts, sesame seeds, almonds, etc.

Dr. Campbell reported to the New York Times that there was basically no osteoporosis in China, yet the calcium intake ranged from 241-943 mg per day (average, 544). The comparable U.S. calcium intake is 841-1,435 mg per day (average 1,143), mostly from dairy sources. Osteoporosis and other bone diseases are major public-health concerns in Westernized societies.

Although other epidemiologic studies have linked osteoporosis not only to low calcium intake, but to other various nutritional factors that cause immoderate calcium reserves over time. So, what are the factors that contribute to an increase in urinary calcium excretion?

Dietary Factors that Induce Calcium Loss in the Urine

- Animal Protein
- Salt
- Caffeine
- Refined Sugars
- Nicotine
- Aluminium containing antacids
- Drugs: such as antibiotics, steroids and thyroid hormones
- Vitamin A Supplements

Data published also clearly links increased urinary excretion of calcium with animal-products intake, but not with vegetable-protein intake.[27]

The Nurses' health Study also found that women who consumed 95 grams of protein per day had a 22% greater risk of forearm fractures: than those who consumed less than 68 grams.[28]

But, the most comprehensive epidemiological survey involving hip fractures and food was done in 1992.

The authors of this survey did their homework well. They sought out every peer-reviewed geographical report ever done on hip-fracture incidence. They located thirty-four published studies of women in sixteen countries. Analysis showed that diets high in animal protein had the highest correlation with hip-fracture rates: an 81% correlation between eating animal protein and fractures. The extra calcium contained in dairy foods simply cannot counteract the powerful effects of all the dietary factors that induce calcium loss in the urine.[29] Basically drinking more milk is simply not protective, unless, maybe you completely reduce, or even better, completely avoid taking the other dietary factors than can induce calcium loss.

The Eskimos are a perfect example. They consume a huge amount of calcium, over 2,000 mg per day, from all the soft fish bones they eat. Yet, Eskimos have the highest hip-fracture rate in the world. Why? It's may be because Eskimos consume vast amounts of animal proteins from fish.[30]

Calcium absorption rates, according to the American Journal of Clinical Nutrition:

- Brussels sprouts 68.8 percent
- Mustard greens 57.8 percent
- Broccoli 52.6 percent
- Turnip 51.6 percent
- Kale 50 percent and
- Cow's milk 32 percent

To abstain from dairy products, milk substitutes from alternative food sources are commonly available and easily produced.

"Almond, Rice, Soya, Flaxseed or any other forms of nuts or seeds can be alternative cow's milk choices, but grounded almond and Soy milk are the most common milk substitute sources that are mainly sought-after by vegetarians and vegans today."

Milk *does not* have to come from a cow or from any other mammal.

Compared to cow's milk, almond milk is the much healthier alternative choice of milk. Almond milk contains a wider variety of the vitamins and minerals that your body needs to function. Almond milk also contains higher levels of antioxidants and fibre, overall making it a better substitute liquid source over cow's milk.

How to Make Your Own Almond Milk

1. Soak 1-1½ almonds in water overnight. The soaking softens the Almonds well for use, whilst also making their nutrients bio-available
2. Drain water, clean and rinse almonds thoroughly
3. Blend 1-1½ cups of clean almonds in 3-4 cups of fresh spring or filter/bottled water
4. Using a filter nut-milk-bag: Squeeze and filter the almonds milk into a clean container. The filter will retain the almond pulp, whilst just leaving the almond milk.
5. Store: Almond milk can be stored and kept in the refrigerator for up to 4-6 days, covered. It will usually need shaking before serving, to remix the liquid
6. Use: Almonds milk can be used with a range of foods where you would use dairy milk, such as for cooking, cereals, for drinking, etc.
7. The almond pulp should not be wasted as it can be used to add more fibre to your diet.

Almond milk is one of the most nutritionally valuable milk substitutes available today. It is high in a number of vitamins and minerals: Vitamin E, Vitamin H (biotin), manganese, magnesium, phosphorous, potassium, selenium, iron, zinc and calcium. Almond milk is low in calories, at only 40 calories per eight-ounce serving, and is low in fat. It contains only three grams of fat per eight-ounce serving.

Almond milk is lactose, casein and cholesterol free; it's also free of saturated fats and contains more of the healthy monounsaturated fats.

Notes to think of: Almonds milk is not a highly good source of calcium, so it's important to consume you almond milk with oats, nuts, and seeds, topped up with a selection of nutritious fruit. Also remember

to eat plant foods that contain high absorption rates of calcium – Green-leafy vegetables like broccoli, Brussels sprouts, spinach, kale etc, throughout the day.

Do we Look like Baby Cows?

Let us be realistic about things. Naturally, in nature "cow's milk" was only intentionally designed for baby calves to drink. Like all female mammals (including humans) who produce milk for their young, a female cow secretes milk that is intended to nurse her baby calf for about one year. This fatty hormonal secretion (containing over 60 hormones and growth factors), which are specifically designed to take her 65-pound newborn calf and turn it into a 700-pound cow in less than one year: that is quite a feat!

All animals drink their own milk. Humans are the only mammals that drink other species' milk. So why do humans drink it? That is a good question.

Dr. Neal Barnard confirms the afore-said:

> The amazingly successful and expensive advertising campaigns of the dairy industry not only got our mothers to feed us formula instead of breast milk, but seem to have convinced us that it is "natural" for people to drink cow's milk. Nothing could be less natural. No species drinks milk beyond infancy and none consume the milk of other species.

Nature has provided all mammals with their own "mother's milk". Cow's milk has the essential nutrients and growth hormones for their baby calves. Do we look like baby cows?

Leading pediatricians, such as Dr. Benjamin Spock (also an influential and best-selling author of parenting books) and Dr. Jay Gordon, have been strong proponents of a "dairy-free diet" for children. The Physicians Committee for Responsible Medicine (PCRM), a non-profit organization whose members include 150,000 health professionals, has always been adamantly opposed to cow milk consumption.

Milk today is pasteurized to remove harmful pathogenic micro-organisms, for decreasing bacterial load, for reducing souring and for prolonging milk self-life. However, types of pasteurizations or heat treatments can cause important nutrients to be destroyed, denatured or missing in the final product.

Another problem is that in some countries hormones are still being added to animal feeds or have been purposely injected into animals: especially in cows and chickens, in order to greatly increase milk production and tissue bulk. These added chemicals may evidently lead to infections in cows. Cows may then require additional antibiotics to remove the causative pathogen, in order to make sure cows do not develop "mastitis". If the cow does get mastitis, then antibiotic residues, pus and blood cells are then retained and found in the cow's milk.[31]

During pregnancy, a cow's oestrogen level skyrockets over 30 times higher than when she's not carrying a calf. Cow's milk contains *natural bovine estrogens and other growth factors*, including Insulin-like-Growth Factor-1 (IGF-1). IGF-1 is a growth factor that can increase cancer growth. Cow's milk and other milk food derivatives have been the subject of many epidemiological studies and they have been proved to offer a high risk for women of developing breast cancer.[32]

The Health Concerns of Cow's Milk towards Human Babies and Infants

Babies fed cow's milk in the first year of their lives are also at risk of developing iron deficiency. Many authorities, including the American Academy of Pediatrics on Nutrition, for safety reasons highly recommend that cow's milk should be excluded from the diet in the first year of life.[33] The mineral iron in cow's milk (there is an average of four times more iron in cow's milk than in human breast milk) is not easy for babies to absorb, and cow's milk appears to interfere with the body's absorption of iron from other foods. Worse, cow's milk has been shown to cause iron loss by producing gastrointestinal bleeding.[34]

Pediatricians learned long ago that cow's milk was often a cause of colic in young infants. Breastfeeding mothers can have colicky babies whilst consuming cow's milk, because cow's antibodies can pass through the mother's blood stream into the breast milk to the baby.[35] This may even cause allergies to cow's milk in infants.

Human infants receiving cow's milk protein may be persistently restless and unhappy, screaming at intervals and appearing in pain. Infants may also appear excessively greedy. They may also regurgitate freely and may have loose, mucous stools in which blood and sometimes sugars are detected. Infants, especially if their parents or siblings suffer from eczema or hay fever or asthma, may develop facial or generalized eczema, persistent nasal congestion and noisy wheezing, whenever cow's milk protein is introduced. These symptoms may be accentuated at that time, with or without gastrointestinal problems.[36] Milk is one of the most common causes of food allergies in humans, and it is the single most common cause of allergy in infants.[37]

COW'S MILK IS THE PERFECT FOOD FOR BABY CALVES. IT IS *NOT* HEALTHY FOR HUMANS

Dr. Neal Barnard wrote his book "breaking the food seduction".[38] There actually is a substance found in dairy products that looks very much like morphine called "casomorphin."

A protein is made from a constituency of amino acids (the building blocks of protein). In your digestive tract proteins are naturally broken down into individual or chains of amino acids, which are then used to produce other proteins or glycoprotein's (enzymes, hormones, muscle tissue, collagen, etc). But, the protein casein found in dairy products does not function in the same way.

Casein is a long-chain of amino acids (it's a protein) like other proteins, but in the calf's digestive tract it's broken-down into not individual amino acids, but stings of 4, 5 or 7 amino acids.[39]

Types of Casomorphins	Amino acid sequence
B-casomorphin-7 (bovine)	Try-Pro-Phe-Pro-Gly-Pro-Ile
B-casomorphin-7 (human)	Try-Pro-Phe-Val-Glu-Pro-Ile
B- casomorphin 5 (bovine)	Try-Pro-Phe-Pro-Gly
B- casomorphin 5 (human)	Try-Pro-Phe-Val-Glu
Morphiceptin	Try-Pro-Phe-Pro-NH2

As the above diagram shows, there is also the potential for release of casomorphins from human milk. However, human BCM7 (Tyr-Pro-Phe-Val-Glu-Pro-Ile) differs from the bovine form (Tyr-Pro-Phe-Pro-Gly-Pro-Ile), and so does BCM5, at two amino acid positions.

They are biologically active (they have opiate-activity) and work like mild narcotic drugs. Not that strong, but they do have the narcotic effects of 1/10th the powerful effect of pure morphine. Why are these chemicals found in cow's milk (B-casomorphin -7 and 5), in human breast milk (B-casomorphin 5) and in nature?

The potential release of B-casomorphins varies between species and breeds. Nature has biologically added milk with protein, fat, sugar and growth factors also combined with a nice narcotic affect

(Casomorphins). The mother and infant bond has a biological basis: drug-like compounds called casomorphins are naturally produced for each species. The protein casein found only in milk, breaks down in the stomach to produce a substance called B-casomorphin, which as its name applies, has an opium effect in humans and in other animals. This makes sense from an evolutionary standpoint, as species survival may depend on a "close maternal bond" between infant and mother. Casomorphin's process this opium effect to help heighten the "Mother-Infant Bond."[40]

Cow's Milk Casomorphins Have Been Shown to be Contributory Factors to the Increased Risk of Apnoea

Attacks of apnoea (absence of breathing) and muscular Atony (muscle that has lost its strength) after exposition to cow's milk may be also explained by extra-central activity of Bovine B-casomorphin-7 (bBCM-7). This cow's milk compound may furthermore be responsible for pseudo-allergic reactions resulting from the direct release of histamine from mast cells without IgE.[41] It may be related to degranulation of mast cells and an increase of their tryptase activity, which is identified in infants who died of Sudden Infant Death Syndrome (SIDS), that do not breastfeed. Breastfeeding reduced the risk of sudden infant death syndrome by 50% at all ages throughout infancy.[42]

Bovine Casomorphins Transferring to Babies via Human Milk

Natural casomorphin-5 found in human-breast milk is logically produced in humans and is not considered unhealthy; drinking other opiate-like compounds from the milk of other mammals may produce drastic side-effects to the human body.

Casomorphins liberated from cow's B-casein are accused of participating in the aetiology of problematic conditions: autism, increased mucus production and crib death, which is, the most common cause of death for healthy infants after one month of age. 1/2000th American babies actually die in this way. Every day, six babies' stop crying and six parents sadly start; other symptoms

COW'S MILK IS THE PERFECT FOOD FOR BABY CALVES. IT IS *NOT* HEALTHY FOR HUMANS

of consuming cow's milk casomorphins include Type 1 diabetes; postpartum psychosis; circulatory disorders and food allergies.[43] Whereas human casomorphins (irHCM), which are the only ones found in the breast-milk of women, who do not drink cow's milk are associated with normal psychomotor development and muscle tone. In contrast, elevated basal levels of Bovine casomorphins (irBCM) found in cow's milk based formula-fed-infants for example, was associated

in a delay in psychomotor development and heightening muscle tone spasticity. This overall evidence clearly suggests that the inability of some infants to adequately eliminate *"Bovine-casomorphins"* maybe a risk factor for delay in psychomotor development and other diseases, such as autism.[44]

The Link Between Cow's Milk and Diabetes

Other unhealthy consequences of dairy products have been linked to insulin-dependent diabetes mellitus (IDDM) or childhood-onset. Epidemiological studies of various countries show a strong correlation between the use of dairy products and the incidence of IDDM.[45]

This disease tends to strike in early teenage years and accounts for many deaths worldwide. IDDM starts by the immune system attacking and destroying the beta cells of the pancreas which produce insulin. As already mentioned, there is normally a genetic predisposition, but mounting evidence suggests that the disease is linked to an allergy to bovine (cow's) serum protein that may be seen as foreign to the human immune system. Human autoantibodies are produced by the B-cells of the immune system, due to the ingestion of foreign bovine proteins; which start attacking the beta-cells of the pancreas that produce insulin, causing IDDM and other degenerative diseases.

Milk is designed to be a highly concentrated source of nutrients, which ensure the growth and well-being of new-borns and young mammals. Sadly, man-made chemicals, which the body cannot distinguish from natural substances, may also become concentrated in milk by the same process. The carcinogenic industrialized contaminants in milk (DDT, fat-soluble PCB's and dioxins) have been reviewed, and government action has been taken to try to reduce these pollutants in milk, but without much success.

Toxins and carcinogens are deposited and accumulate in adipose tissues (adipocytes) of animals. Whenever we eat animal flesh, we also have the tendency of accumulating these harmful toxins. These and other pollutants, which can cause cancer and which are also powerful endocrine disrupters, can be particularly concentrated in milk from pasture or feed, together with some radioactive isotopes released as a result of nuclear accidents. Recent studies have shown that breast milk from "clean human beings" has up to 350 artificial chemicals or pollutants in it.[46]

Many hormones are naturally found in mammal milk: prolactin, oxytocin, adrenal and ovarian steroids, insulin, relaxin, calcitonin, somatostatin, neurotensin and prostaglandins. These substances are all found in greater levels in cow's milk than in breast-feeding mother's blood: together with thyroid-stimulating hormone (TSH), thyrotrophic-realising hormone (TRH), thyroxine, erythropoietin and bombesin – although these substances are at lower levels in milk than in maternal blood. Milk also contains numerous growth factors that include: IGF-1, epidermal growth factor (EGF) and nerve-growth factor (NGF), including over 40 enzymes that help the development of the infant's immunological function, and in some cases the maturation of specific cells and tissues.[47]

All mature breast milk, whether from humans or other mammals, is a medium for transporting abundant amounts of distinctive chemical components, and these components vary in composition: between species, between mothers, between feeds and during the course of lactation.[48] Different breast teats have even been shown to produce milk of different composition – to suckle different young animals – with different nutritional needs.

The point is this! All mammalian milk, whether from humans, cows or other species, is a powerful biochemical solution of enormous complexity, uniquely designed to provide for the individual needs of young mammals of the same species. It is not that cow's milk is not a good food. Cow's milk is a great food – for baby cows! And therein lays the source of the problem. Cow's milk contains abundant amounts of growth-stimulating hormones for baby cows to grow and these are not intended for humans: Babies, infants, children or human adults to drink.

Too much of these hormones can cause havoc to the human body, and may cause many health concerns: during childbirth, infancy and all through human life. If the mother, after breast-feeding her child, decides to give cow's milk to her infant, as scientific literature proves, severe health consequences will arise and disease can develop in the child and in all those persons who continue to consume cow's milk in their adult years. Cow's milk should NEVER be drunk by humans, neither at birth nor during adulthood. Never! Ever! Human breast milk is best for the first 6 months of human life or possibly longer, if the mother can produce enough milk for her newborn.

As you have read and the many studies and reviews have clearly indicated, cow's milk should only be drunk by baby cows because it was naturally designed for cows to drink. Pregnant mothers, who choose to drink cow's milk before, in-between, or after pregnancy, are increasing their own risk, and exposing their babies and their families to these many ailments. What is the best milk to drink? Human breast milk until weaning, then other sources of natural-food liquids should be consumed. E.g. water, almond milk, rice milk and other types of alternative liquids. These types of liquid foods can also be combined and mixed (bended) with healthy nutritious fruits.[49]

Human breast milk is the best and only source of natural milk. By also eating a rich, dense, nutritious, anti-cancer, plant-based diet that contains powerful, protective nutrients, we can also reduce the accumulation of harmful man-made chemicals and other harmful carcinogens: to detoxify the body, boost our immune system and other vital organs. These actions, without any doubt, will reduce our risk of developing acute and chronic-degenerative diseases.[50]

We need to turn the tide against the milk pandemic and stop drinking mammalian milk from other species, except our own mother's-breast milk, until weaning.

PROFESSOR JANE A. PLANT:
HER FIGHT AGAINST BREAST CANCER

An inspirational true story

I would like to personally thank Dr. Jane Plant for her acknowledgment and consent in letting me recount her amazing story. Her inspiring true story will certainly give hope to those women who already have breast cancer or want to protect themselves from this disease which mainly affects women and a small majority of men. This is a story that proves that by changing your dietary habits and lifestyle you can ultimately help reduce your risk of developing breast cancer and you may even cure yourself from this disease if you already have it.

Dr. Jane Plant was diagnosed with breast cancer. She had a radical mastectomy, 3 further operations, 40 radiation treatments and 12 chemotherapy treatments. None of these conventional treatments worked. The cancer kept coming back. Not just once, twice, but five times, and to different parts of her body.

The last tumour metastasized to her neck, and chemotherapy was not effective. She then adopted a plant-based diet and eliminated all dairy products from her diet. Within days the lump started to shrink. Two weeks later, after her second chemotherapy session and one week after she gave up dairy produce, the lump in her neck started to itch, soften and shrink. Within 6 weeks the tumour was completely gone. She had cured herself of her disease.

The first clue to understanding what was promoting Jane's breast cancer came when her husband Peter, who was also a scientist, returned from China at that time when she was undergoing another chemotherapy session.

He had brought with him letters and cards, as well as some herbal suppositories, sent by her friends and scientist colleagues in China. The suppositories were sent to her as a cure for breast cancer. Despite the awfulness of the situation, they both had a good laugh, and she remembered saying that if this was the treatment for breast cancer in China, it was little wonder that Chinese women avoided getting the disease.

Why did Chinese women not get breast cancer? Jane had graduated as a geochemist. She had collaborated on several studies with Chinese colleagues on the links between soil chemistry and disease, and she remembered some of these statistics.

Breast cancer in China was very rare throughout the whole country. At the time only 1 in 10,000 women in China had breast cancer, compared to that terrible figure of 1 in 12 in Britain, and the even grimmer average of 1 in 10 across most Western countries. It is not just a matter of China being a more rural country, with less urban pollution. In highly urbanized Hong Kong, the rate staggered to 34 women in every 10,000, but this still puts most Western countries at higher rates of mortality.

Something sparked inside Jane's head. It became obvious to her that some lifestyle factor, not related to pollution, urbanization or the environment, could be the cause that was seriously increasing Western women's risk to the development of breast cancer.

Jane then discovered that whatever is causing the huge differences in breast cancer rates between oriental and Western countries was surely not genetically linked. Scientific research showed that whenever Japanese or Chinese people migrated to the West, within one or two generations their rates of breast cancer dramatically increased.

The same thing happens when oriental people adopt a completely Western lifestyle in Hong Kong. The slang name for breast cancer in China translates as "rich woman's disease"! This is because in China only the better-off can afford to eat what is termed "Hong Kong food" or "junk food".

Breast cancer is a "middle-class disease" that attacks primarily the wealthier and higher socio-economic groups, those that can afford to eat rich foods. But, where were fats and hormones coming from in the diets of Western countries, i.e. meat, eggs and mostly in dairy products? Did Chinese people consume a lot of animal products, especially dairy? No, they did not.

Before Jane had breast cancer for the first time, she had eaten a lot of dairy produce, such as skimmed milk, low-fat cheese and yoghurt. She had used it as her main source of protein. Jane also ate cheap, but lean, minced beef, which she now realized was probably often ground-up dairy cow, which later proved to be another risk factor for cancer. While she had been ill Jane had been eating only one low-fat yoghurt per day. With her husband Peter's and her own insight into the Chinese diet, Jane decided to give up all types of dairy products. She was amazed how many products, including commercial soups, biscuits and cakes, contain some form of dairy products.

Jane now believes that there is a link between dairy products and breast cancer. She believes that identifying the link between breast cancer and dairy produce, and then developing a diet specifically targeted at maintaining the health of her mammary and hormonal system, had cured her of her own breast cancer.

Jane believes it is just unnatural to drink milk from another species of animal, and I totally agree.

Jane now knows that increased stress-relative events, hormones and specific foods (especially dairy products and animal products) can increase human risks of breast, prostate, colon, testicular and ovarian cancer.

Jane is still, to this day, breast cancer-free and dairy-free.

Notes for chapter 24

1   *Say No WAY! To WHEY!* Robert Cohen Executive Director and author of Milk A-Z. Retreived from http://www.notmilk.com/whey.html

2   The Lancet. (1974), **2**:1061; Neuroepidemiology (1992), 11:304A¬12.

3   Cohen, "Say No Way!"

4   Bertron, P., Barnard, N.D., & Mills, M. (1999). Racial Bias in Federal Nutrition Policy, Part I: The public health implications of variations in lactase persistence. *Journal of the National Medical Association*, **91**, pp.151-157.

5   Cramer, D.W *et al.* (1989). Galactose consumption and metabolism in relation to the risk of ovarian cancer. *The Lancet*, **2**, pp.66-71.

6   Plant, J.A. (2007). *Your Life In Your Hands. Chapter 3: The Third Strawberry*, pp.88-89. Published by Virgin Books. Cramer, D.W. (1989). Lactase persistence and milk consumption as determinants of ovarian cancer risk. *American Journal of Epidemiology*, **130**, pp.904-910.

7   Adami, H.O., Hunter, D., & Trichopoulos, D. (2002). *Textbook of Cancer Epidemiology*. New York. Published by Oxford University Press. Kaufman, F.R *et al.* (1981). Hypergonadotropic hypogonadism in female patients with galactosemia. *New England Journal of Medicine*, **304**(17), pp.994-998.

8   Iacono, G. *et al.* (1998). Intolerance of cow's milk and chronic constipation in children. *New England Journal of Medicine*, **339**, pp.110-114.

9   Scott, F.W. (1990). Cow milk and insulin-dependent diabetes mellitus: is there a relationship? *American Journal of Clinical Nutrition*, **51**, pp.489-91. Knip, M *et al.* (1992). A bovine albumin peptide as a possible trigger of insulin-dependent diabetes mellitus. *New England Journal of Medicine*, **327**, pp.302-307.

10 Feskanich, D., Willet, W.C., Stampfer, M.J., & Colditz, G.A. (1997). Milk, dietary calcium, and bone fractures in women: a 12-year prospective study. *American Journal for Public Health*, **87**, pp.992-997.

11 Pennington, J.A.T. (2009). Bowes and Churches Food Values of Portions Commonly Used. (19th edition). Published by LWW. Ornish, D *et al.* (1990). Can lifestyle changes reverse coronary heart disease? *Lancet*, **336**, pp.129-133.

12 Knip. M *et al.* (1999). Cow's Milk Formula feeding induces primary immunization to insulin in infants at genetic risk for type 1 diabetes. *Diabetes*, **48**(7), 1389-1394.

13 Barnard, N *et al.* (1997). Dairy products and breast cancer: the IGF-1, estrogen, and bGH hypothesis. *Medical Hypothesis*, **48**, pp.453-461; Chan, J.M *et al.* (1998). Plasma insulin-

like growth factor-1 and prostate cancer risk: a prospective study. *Science*, **279**, pp.563-565. World Cancer Research Fund. (1997). *Food, Nutrition, and the Prevention of Cancer: A Global Perspective.* American Institute of Cancer Research. Washington, D.C. Cadogan, J. et al. (1997). Milk intake and bone mineral acquisition in adolescent girls: randomised, controlled intervention trial. *BMJ*, **315**, pp.1255-1269.

14 Chan, J.M et al. (1998a) What causes prostate cancer? A brief summary of the epidemiology. *Semester in Cancer Biology*, **8**, pp.263-273. Armstrong, B, Doll, R. (1975). Environmental factors and cancer incidence and mortality in different countries, with special reference to dietary practices. *International Journal of Cancer*, **15**, pp.617-631. Rose, D.P et al. (1986). International comparisons of mortality rates for cancer of the breast, ovary, prostate, and colon, and per capita food consumption. *Cancer*, **58**, pp.2363-2371. Jin, F et al. (1972-1989). Incidence trends for cancer of the breast, ovary, and corpus uterin in urban Shanghai. *Cancer Causes Control*, **4**, pp.355-360. Nagata, C., Kawakami, N., & Shimizu, H. (1997). Trends in the incidence rate and risk factors for breast cancer in Japan. *Breast Cancer Res Treat*, **44**, pp.75-82. Leung, G.M et al. (2002). Trends in breast cancer incidence in Hong Kong between 1973 and 1999: an age-period-cohort analysis. *Br J Cancer*, **87**, pp.982-988.

15 Breslau, N.A et al. (1988). Relationship of animal-protein-rich diet to kidney stone formation and calcium metabolism. *J. Clinical Endocrinology*, **66**, pp.140-146.

16 Abelow, et al. (1992). Cross-cultural association between dietary animal protein and hip fracture: a hypothesis. *Calcification Tissue International*, **50**, pp.14-18. Feskanich, D et al. (1996). Protein consumption and bone fractures in women. *American Journal of Epidemiology*, **143**, pp.472-479.

17 Lawrence, R.A. (1980). Breastfeeding: A guide for the medical profession. Published by Mosby. Rose, D.P et al. (1986). International comparisons of mortality rates for cancer of the breast, ovary, prostate, and colon, and per capita food consumption. *Cancer*, **58**, pp.2363-71. Talamini, R., La Vecchia, C., Decarli, A et al. (1984). Social factors, diet and breast cancer in a northern Italian population. *British Journal of Cancer*, **49**, pp.723-729. La Vecchia, C., Pampallona, S. (1986). Age at first birth, dietary practices and breast cancer mortality in various Italian regions. *Oncology*, **43**, pp.1-6. Boyd, N.F et al. (1993). A meta-analysis of studies of dietary fat and breast cancer risk. *British Journal of Cancer*, **68**, pp.627-636. Levi, F et al. (1993). Dietary factors and breast cancer risk in Vaud, Switzerland. *Nutrition and Cancer*, **19**, pp.327-235; Gaard, M et al. (1995). Dietary fat and the risk of breast cancer: a prospective study of 25,892 Norwegian women. *International Journal of Cancer*, **63**, pp.13-17.

18 McDougall, J. (1985). *McDougall's Medicine* (Piscataway N). New Century Publishers, p.67.

19 Anderson, M. (2009). Healing cancer from inside out. Part 3: The Rave diet and lifestyle: Calcium Needs, Osteoporosis and Acidic Diets, pp. 139-141. Published by www.RaveDiet.com

20 Mangione, K.K et al. (July 2005). Exercise Prescription for a patient 3 months after hip fracture. *Physical Therapy*, **85**(7), pp.676-687. Young Y, Brant L, German P et al. (1997). A longitudinal examination of functional recovery among older people with subcapital hip fractures. *J Am Geriatr Soc*, **45**, pp.288–294. Aloia, J. (1981). Exercise and skeletal health. *Journal of American Geriatric Society*, **29**(104). Smith, E et al. (1981). Physical activity and calcium modalities for bone mineral increase in aged women. Med Sci, Sports Exercise, **13**(1), pp. 60-64. Smith, E,L.(1982). Exercise for prevention of osteoporosis: A review. *The Physician and Sportsmedicine*, **10**(3), pp.72-82.

21 Ibid.

22 Ibid.

23 Pritikin, N. (1984). The Pritikin Program for Diet & Exercise, p. 44. Published by Bantam.

24 McDougall's Medicine, John A. McDougall, M.D., p.70. Wohl, G.M et al. (1974). Modern Nutrition in Health and Disease (5th ed), p.274. Published by Lea and Febiger. C. Paterson. (1978). Calcium Requirements in man: A critical review. *Postgrad Med J*, **54**, pp.244.

25 Maggi S et al.(1991). Incidence of hip fractures in the elderly: a cross-national analysis. *Osteoporosis International*, **1**, pp.232-241.

26 Feskanich, D et al. (1997). Milk, dietary calcium, and bone fractures in woman: a 12-year prospective study. *American Journal of Public Health*, **87**, pp.992-997.

27 Sellmeyer DE et al. (January 2001). A high ratio of dietary animal to vegetable protein increases the rate of bone lose and the risk of fracture in postmenopausal women: Study of Osteoporotic Fractures Research Group. *American Journal of Clinical Nutrition*, **73**(1), pp.118-122.

28 Jajoo, R et al. (June 2006). Dietary acid-base balance, bone resorption, and calcium excretion: *Journal of the American Coll Nutrition*, **25**(3), pp.224-230.

29 Abelow, BJ et al. (1992). Cross-cultural association between dietary animal protein and hip fracture: A hypothesis. *Calcification Tissue Int*, **50**(1), pp.14 -18.

30 Mazess RB et al. (1997). Bone mineral content of North Alaskan Eskimos. *American Journal of Clinical Nutrition*, **27**(9), pp.916-925. Pawson IG. (1974). Radiographic determination of excessive bone loss in Alaskan Eskimos. *Human Biology*, **46**(3), pp.369-380.

31 Anderson, J. *Mad Poisoned Cows-The Dangerous Toxins in Meat and Milk.* www.alternativemedicine.com

32 Barnard, N. (1996). Milk and Breast Cancer. *Prevention and Nutrition*, pp.11-17. Anderson, J., ibid.; Kradjian, R. *The Milk Letter: A message to my patients*. Retreived from http://www.notmilk.com/kradjian.html

33 American Academy of Paediatrics Committee on Nutrition (1992). The use of whole cow's milk in infancy. *Paediatrics*, **89**, pp.1105-1109. Plant, J.A. (2012). *Your Life in Your Hands: Chapter 4: Rich Woman's Disease*, pp.97-98. Published by Virgin Digital; Revised edition.

34 Anyon, C.P., & Clarkson, K.G. (1971). A cause of Iron Deficiency anaemia in infants. *New Zealand Medical Journal*, **74**, pp.24-25.

35 Clyne, P.S & Kulczycki, A. (1991). Human breast milk contains bovine IgG. Relationship to infant colic? *Paediatrics*, **87**(4), pp.439-444. Wilson, J.F., Lahey, M.E. & Heiner, D.C. (1974). Studies on iron metabolism. Further observations on cow's milk-induced gastrointestinal bleeding in infants with Iron-deficiency anaemia. *Journal of Paediatrics*, **84**, pp.335-344.

36 Ibid.

37 Ibid.

38 Neal Barnard, M.D. Chocolate, Cheese, Meat, and Sugar: Physically Addictive; Source from http://www.youtube.com/watch?v=5VWi6dXCT7I youtube video.

39 Ibid.

40 Ibid.

41 Neal Barnard. (2004). *Breaking the Food Seduction: The Hidden Reasons Behind Food Cravings –And 7 Steps to End Them Naturally*. Published by St. Martin's Griffin; Reprint edition.

42 Kurek M *et al*. (1992). A naturally occurring opioid peptide from cow's milk, beta-casomorphine-7, is a direct histamine releaser in man. *International Archive Allergy Immunology*, **97**(2), pp.115-20.

43 Vennemann, M.M *et al*. (2009). Does breastfeeding reduce the risk of Sudden Infant Death Syndrome? *Journal of Paediatrics*, **123**(3), pp.406-410.

44 Wasilewska, Jolanta *et al*. (June 2011). Cow's-milk–induced Infant Apnoea with Increased Serum Content of Bovine β-Casomorphin-5. *Journal of Paediatric Gastroenterology & Nutrition*, **52**(6), pp.772–775. Kurek, "A naturally occurring opioid peptide from cow's milk, beta-casomorphine-7", 2009.

45 Scott, F.W *et al*. (1990). Cow milk and insulin-dependent diabetes mellitus: is there a relationship? *American Journal of Clinical Nutrition*, **51**, pp.489-491.

46 The Guardian, 13th July, 1999.

47 Akre, J. (1989) Infant feeding: the physiological basis. *WHO Bulletin, OMS Supplement,* **67**(Suppl), pp.9-18. Lawrence, Ruth A (1980). *Breastfeeding: A guide for the medical profession.* Published by Mosby.

48 Ibid.

49 Feskanich, D *et al.* (1996). Protein consumption and bone fractures in women. *American Journal of Epidemiology,* **143**, pp.472-479.

50 Ibid.

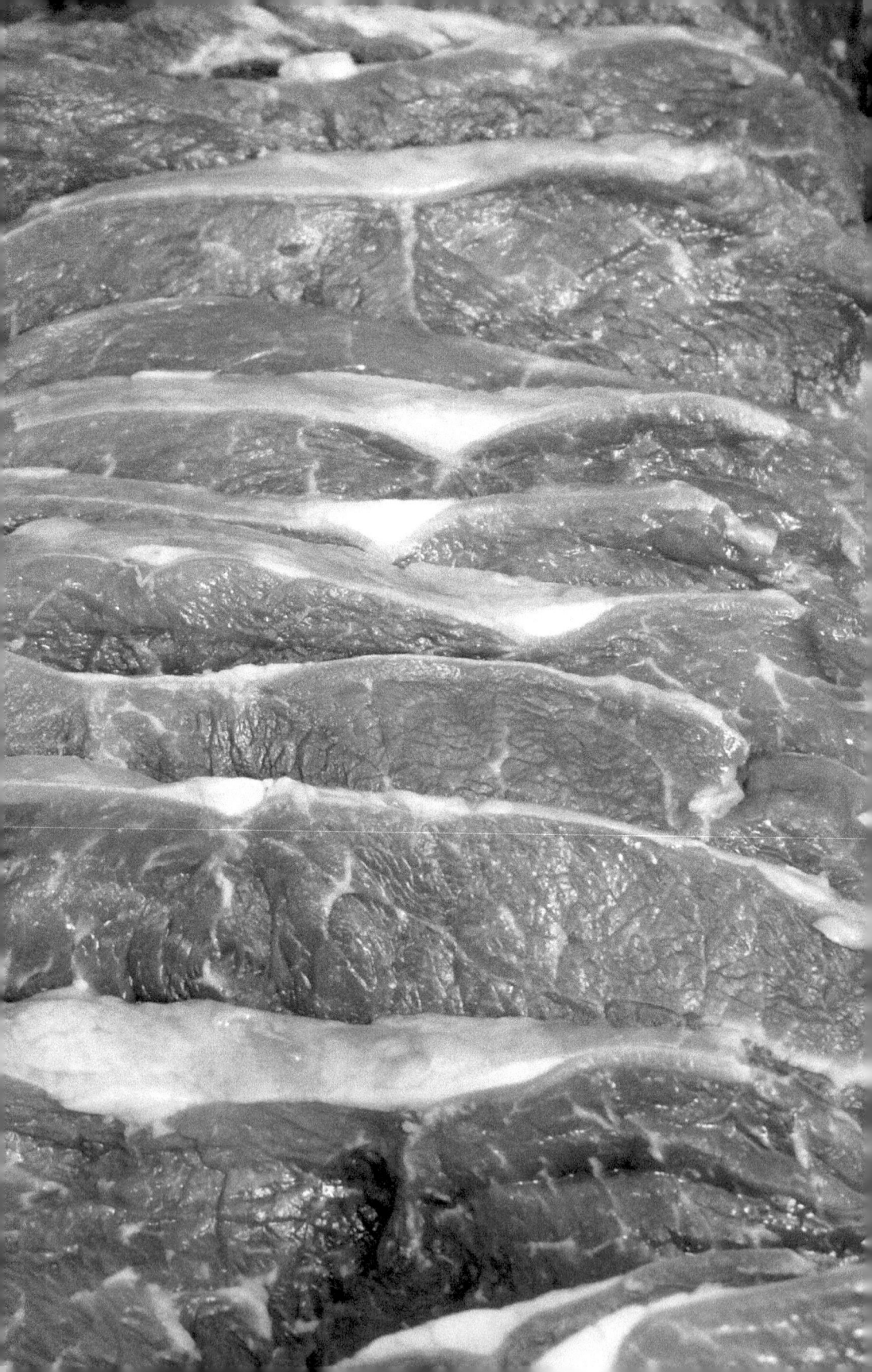

25

Animal Products: The Slow-Growing, Fleshy-Road to Chronic Diseases

The beef industry has contributed to more American deaths than all the wars of this century, all natural disasters, and all automobile accidents combined. If beef is your idea of "real food for real people" you'd better live real close to a real good hospital.

Neal Barnard, M.D.
Founder and president of the Physicians Committee for Responsible Medicine (PCRM)

For many persons the pleasure of eating animal products has a lot of downsides, especially for their health. Many still think they will not get enough calcium or protein in their diet if they completely eliminate dairy and other animal products. I consider this a tradition that has slowly become a cultural eating style.

Societies today have been brainwashed by TV advertisements, sponsored and paid by the meat and dairy industries, which have succeeded to sell their products by persuading the general public that they need to eat animal products to sustain optimum health.

This is certainly not true. You can get all your adequate recommended daily dietary needs of protein and calcium by eating a variety of plant foods. As you will read, and the scientific studies have revealed, rich sources of animal proteins can also be problematic to human health. As we have already discussed in previous chapters, animal products do not contain the powerful-protected compounds called phytochemicals or many other essential micronutrients, including dietary fibre, which are significantly crucial to human health and disease prevention.

In the China project, Dr. T. Colin Campbell wanted to know whether or not, and if yes, which types of proteins promote cancer growth. We now know that chemical carcinogens are the seeds that initiate cancer, but do not promote the disease unless the environmental and nutritional conditions are right. It is highly probable that for much of our daily life we are being exposed to small amounts of cancer-causing agents.

The 3 Stages of Cancer[1]

Cancer, as we know it, proceeds through three stages: initiation, promotion and progression. As a basic analogy, cancer process is similar to planning a lawn.

Initiation is when you plant the seeds in the soil, promotion is when the grass starts to grow, and progression is when the grass grows completely out of control – invading the driveway, the shrubbery, and the sidewalk.

ANIMAL PRODUCTS: THE SLOW-GROWING, FLESHY ROAD TO CHRONIC DISEASES

So which process "implants" the grass seeds in the soil in the first place? Chemicals that do this are called "carcinogens." The chemicals are most often the by-products of industrial processes, although small amounts can be formed in nature as in the case with aflatoxin, or are self-inflicted like the carcinogens in cigarette smoke. These carcinogens may genetically transform or mutate normal cells into cancer-prone cells. A mutation involves permanent alteration of the genes of the cell, with damage to its DNA.

The complete initiation stage can develop in a very short period of time, even within minutes. It is the time required for the chemical carcinogen to be consumed, absorbed into the blood, transported into cells, changed into its active product, bound to DNA, and passed on to the "daughter cells".

When the new daughter cells are formed, the process will be complete, and the genetic damage may give rise to cancer. The seeds have been put in the soil, ready to germinate. Initiation is complete.

Although the initiated cells are not considered to be reversible, the cells growing through the promotion stage are usually considered to be reversible. This is a very exciting concept. This is the stage that especially responds to nutritional factors. For example, the nutrients from animal-based foods, especially animal protein, promote the development of the cancer, whereas the nutrients from plant-based foods, especially the antioxidants, reverse the promotion stage. This is a very promising observation, because cancer proceeds forward or backward as a function of the balance of promoting or anti-promoting factors found in the diet. Thus, consuming anti-promoting plant-based foods tend to keep the cancer from going forward, perhaps even reversing the promotion. The difference between individuals almost entirely relates to their dietary habits and lifestyle practices.[2]

The second stage of growth is called *promotion*. Like seeds ready to sprout blades of grass and turn into green lawn, our newly-formed cancer-prone cells are ready to grow and multiply, until they become a visible and detectable cancer. The promotion stage does indeed take a far longer period of time than initiation, often many years for humans, until clusters of damaged cells grow into larger and larger masses, and thus a clinically visible tumour is formed.

Like seeds in the soil, the initial cancer cells will not grow and multiply, unless the right conditions are met. For example, the seeds in the soil need a healthy amount of water, sunlight and other nutrients before the grass turns into a fully-grown lawn.

If any of these conditions are lacking, the seeds will not grow. If any of these factors are denied after growth starts, the new seedlings will become dormant, awaiting further supplies of the missing factors. This is the most profound feature of the promotion stage of cancer. Promotion is reversible, depending on whether the early cancer growth is given the right conditions in which to grow.[3]

This is where certain dietary factors become so important. These dietary factors, called promoters, feed cancer growth (as already mentioned in an earlier chapter). Other dietary factors, called anti-promoters, can slow or alter cancer growth. Cancer growth flourishes whenever there are more promoters than anti-promoters, whenever anti-promoters prevail, cancer growth slows or stops. It could be considered as a push-pull process.

The third stage, called *progression*, begins when a bunch of advanced cancer cells progress in their growth, until they have done their final damage, like a fully-grown lawn invading everything around it.

Evidently, the tumor can also invade neighboring or distant tissues. When cancer takes on these deadly properties, it is considered malignant. It can then break away from its initial home and wanders. This process is called metastasizing. This final stage of cancer, if not controlled accordingly, and the final outcome may result to human death.

Animal protein, especially the protein in cow's milk "casein" has been studied as a promoter of cancer.

Animal protein tends to be a promoter of cancer, whilst plant protein does not: indeed, it may actually protect the human body from developing cancer.

We need not eat animal foods. Animal foods are actually detrimental to the human body, and can cause devastating harm to the overall environment, including pain, suffering and finally death towards unnecessary slaughtered-domesticated animals.

Animal protein	Plant protein
Cancer promoter	Cancer protector
Promotes bone loss	Promotes bone strength
Raises cholesterol	Lowers cholesterol
Raises IGF-1 levels	No effect (accept high Soy Intake)
Accelerates aging	No effect

The Milk protein Casein:
Not all Proteins are Alike in Triggering Cancer Growth

In *The China Study*, Dr. Campbell performed many experiments to see which types of proteins may or may not trigger cancer growth.

One can ask whether it makes any difference between what types of proteins were examine in his experiments? Actually, in some of his experiments Campbell used the protein casein (which makes up 87% of cow's milk protein) to establish whether there is a link between the protein casein and cancer development.[4]

To test his theory about dairy, Campbell fed two groups of rats' diets with different amounts of casein. To initiate cancer development, Campbell fed rats a highly-toxic liver carcinogenic substance called "aflatoxin."

After 12 weeks, all the rats eating a diet with 20% casein had a greatly increased level of early cancer tumour growth: rats eating a 5% casein diet showed no evidence of cancer whatsoever. In switching the rats' diets back and forth – between 5% and 20% casein – Campbell and his researchers observed that whenever the rats were fed 20% casein, early liver tumour growth exploded, while when the same rats were given 5% casein, tumour growth actually went down. Campbell concluded

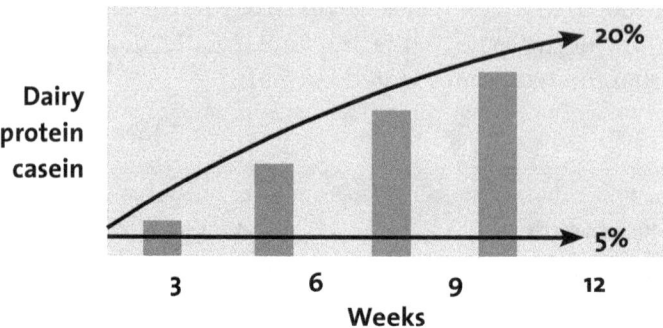

that we could turn on and turn off cancer growth –just by adjusting the right level of intake of the dairy protein "casein".

So the next logical question was whether plant protein, tested in the same way, had the same effect on cancer promotion as casein? The indisputable answer is an emphatic "No!" In Campbell's experiments plant protein did not promote cancer growth, even at the highest levels of intake. Gluten, the protein of wheat, did not produce the same result as casein, even when fed at the same 20% level. Soya was also tested, and the same results were obtained: no cancer growth with 20% soya protein.[5]

The milk protein (casein) in this case was not looking so good. Like flipping a light switch on and off Campbell could control cancer promotion merely by changing the levels of animal protein: regardless of initial carcinogen exposure.

These results were indeed spectacular. Rats generally live for about two years, thus the study was 100 weeks in length. But all animals that were given aflatoxin and fed the regular 20% level of casein either were dead or near death from tumors at 100 weeks. On the other hand, all animals (not only rats on which the study was conducted), that were given the same level of alfatoxin, but fed the low 5% protein diet, were alive and active, with sleek hair coats at 100 weeks. This was a virtual 100 to 0 score, something almost never seen in research before.[6]

At the University of Illinois Medical Centre in Chicago, another research group was working with mammary (breast) cancer in rats. This research showed that increasing intakes of casein promoted the development of mammary (breast) cancer. They found that higher casein intake:

1. Promotes breast cancer in rats dosed with two experimental carcinogens (7, 12-dimethylbenz (a) anthracene (DBMA) and N-nithroso-methylurea (NMU).
2. Operates through a network of reactions that combine to increase cancer; and
3. Functions through the same female hormone system that operates in humans.

Casein affects the way cells interact with carcinogens, the way DNA reacts with carcinogens, and the way cancerous cells grow. The depth and consistency of these findings strongly suggest that they are relevant for humans, for four reasons:

1. Rats and humans have an almost identical need for protein.
2. Protein operates in humans virtually the same way it does in rats.
3. The level of protein intake causing tumour growth in rats is the same level that humans consume.
4. In both rodents and humans, the initiation stage is far less important than the promotion stage of cancer. This is because we are very likely "dosed" with a certain amount of carcinogens in our everyday life, but whether these develop into full-blown tumours depends on their promotion, or lack thereof.

It was now becoming questionable and apparently more obvious, that the human body can develop cancer not by the carcinogen that initiates cancer, but, more evidently, from the nutrient factor or other agents that help promote it. The body has the amazing ability to eradicate carcinogens, but if we eat the wrong foods the initiation step can eventually lead to the promotion stage of tumour-cell development by a specific dietary component. If this dietary component or agent continues to fertilize the malignant tumour, this can lead to an advanced tumour cell growth: the progression stage (metastasis or spread) of the disease.

A pattern was beginning to emerge: the carcinogen was the initiator, while food was the promoter of cancer. Nutrients from animal-based foods *increase* tumour development, while nutrients from plant-based foods *decrease* tumor growth.

Is All This Quackery?

The meat and dairy industry and other medical scientists have considered these findings as quackery (quackery is pretentious talk without sound knowledge of the subject discussed; or deliberate deception; or the promotion, for profit, of a medical remedy known to be false or unproven). There is none of this here.

I personally believe Dr. T. Colin Campbell. There is no "quackery" involved in his experiments on whether there **is** a correlation between animal protein and cancer.

Dr. Campbell did not start with preconceived ideas, philosophical or otherwise, to prove the worthiness of plant-based diets. Indeed, Campbell started at the opposite end of the spectrum. He was a meat-loving dairy farmer in his personal life and an "establishment" scientist in his professional life. Campbell even used to lament the views of vegetarians as he taught nutritional biochemistry to pre-med students.

Campbell's was primarily interested in trying to explain the scientific basis for his views in the clearest way possible. Changing dietary practice will occur and will be maintained only when people believe the evidence and experience the benefits. People decide what to eat for a number of reasons, with health considerations being just one. Campbell's task was only to present the scientific evidence in a form that could be fully proven and clearly understood. The rest is up to us.

Yes! I believe that the protein "casein" found in cow's milk can actually promote cancer growth in humans.

ANIMAL PRODUCTS: THE SLOW-GROWING, FLESHY ROAD TO CHRONIC DISEASES

And, yes! One of our major problems in the so-called developed, rich, industrialized nations is that they consume too much animal protein, which is one risk factor for cancer.

People sometimes ask, "If one is vegetarian or vegan, does one not need an adequate amount of protein?" Yes, of course! Everybody does.

By the way, it is good to know that there are now approximately 6% of the world's population (400 million) who are vegetarians and vegans, and the numbers appear to be growing.

Another frequent question: "Do we need to eat meat for our daily protein needs?"

Let me ask you a question! If you were a cow, where would you do get your daily protein needs? It is from the grass that cows eat. But what about other mammals that eat plants and grass?

Do cows, horses, elephants, rhinos, zebras, giraffes, moose (largest member of the deer family), buffalos and gorillas (largest of all

primates) all eat plant vegetation? Yes, of course, they do, and they are not protein-deficient. And to make this point even clearer, are vegetarians or vegan protein-deficient? No! They are certainly not!

But, most of us do not eat or juice grass (wheat or barley), so what do we need to eat if we want to obtain an adequate and healthy protein source? Answer: a plant-based diet!

Let us go deeper into this subject about protein. Which creatures need the most protein? Babies! Babies have to grow, and they have to grow fast. They weigh around 6-9 pounds when born, and within 3½ months they double their weight: by the end of their first year they should weigh around 20 pounds, i.e. tripling their weight. Don't they need proteins? Where do they get it from? They get it from their own mother's breast milk.

How much protein does mother's milk (human breast milk) have? It is around 1.8%. That is right! Very little protein is found in human breast milk or other mammalian milk.

- How much protein does carrot juice have? Around 1.4%.
- How much protein does baked potato have? Around 1.8%.
- How much protein does oatmeal have? About the same.
- Most plant foods contain around 2-8% plant protein.

Some people think of meat as their main source of protein. But it is easy to get plenty of protein without the fat, the cholesterol, and the other undesirables in meat from other sources: Beans, vegetables, lentils and whole grains, in particular, have more than enough protein to sustain a health-promoting diet.

There are plant foods that contain a larger percentage of proteins: Soy beans, broccoli, kale, spinach, Brussels sprouts, lentils and legumes, etc. Now we know that consuming these foods, which evidently do contain a higher percentage of proteins, are in fact beneficial to human health.

According to Dr. Neal Barnard:

> A variety of grains, legumes, and vegetables can provide all the essential amino acids our bodies require. It was once thought that various

plant foods have to be eaten together to get their full protein value, otherwise known as protein-combining or protein-complementing. We now know that intentional combining is not necessary to obtain all the essential amino acids. As long as the diet contains a variety of grains, legumes and vegetables, protein needs are easily met.[7]

Protein requirements for normal human health should be approximately 5-10% of your daily dietary needs. No more is necessary – indeed, if more than 20% are obtained from animal products this can lead to an increased risk of cancer.

So, eat a variety of plant foods and *you will* get all the essential and adequate amounts of amino acids for all your daily protein requirements.

The Protein Jigsaw-Puzzle: Animal Protein vs. Plant Protein

Firstly, we need to separate fact from fiction, so that we can determine which protein sources our bodies benefit the most from.

In the USA and other Western societies there is the mistaken idea that we need meat, dairy and eggs in order to be properly nourished. Our mind-sets have been programmed with incorrect and dangerous information.

Almost all Americans and Western societies get more than enough of their protein everyday dietary needs. In fact, the average American consumes over 100 grams of protein: about 50% more than the recommended daily amount. However, too many of us, including fitness enthusiasts, athletes, bodybuilders, dieters, and the overweight, turn to protein powders, protein drinks, and nutrition bars in a quest to consume more protein. We should consume around 0.9 grams of protein per body-weight in kilograms.

Example: An 80 kg healthy male should consume around 0.9 x 80 = 72 grams of protein daily.

It is altogether true that in certain circumstances (whenever a person is involved in vigorous and regular physical workouts), additional proteins are required. But the increased need for protein should be also proportionate to the increased need for extra calories burned by the exercise. As exercise increases our appetite, we increase our calorie intake accordingly – and our protein intake increases proportionately.

If we meet the increased caloric demand (as a result of heavy exercise) with an ordinary assortment of natural plant foods – vegetables, whole grains, beans, nuts, and seeds – we will get the precise amount of extra good protein and other nutrient goodies. A typical assortment of vegetables, nuts, seeds, beans, lentils and whole grains supplies about 50 grams of protein per 1,000 calories.

It is important to remember that green vegetables are almost 40-50% protein, and when you eat more vegetables, it gives you your protein in a super-immunity and anticancer package. When the additional calories are derived from health-enhancing plant foods, you will get much more that just protein. These foods will also supply your body with a host of antioxidants to protect you against the increased free radicals generated during exercising. Nature planned this pretty well!

We must consider that the maximum muscle mass the human body can typically add in one week, is about one pound. That is the upper limit of the muscle fibre's capacity to make protein into muscle. Any protein beyond that is simply converted to fat. Although athletes have a greater protein requirement than sedentary individuals, this extra is easily obtained through diet. The use of protein supplements is not merely a waste of money; it is also unhealthy.

Excessive protein, especially animal protein, is no small matter. It ages you prematurely and can cause significant harm. The excess protein you are *not* using is not stored by the body as protein: it is converted into fat or is eliminated via the kidneys.

Eliminating excess nitrogen via the urine drains away calcium and other minerals from the bones, and breeds kidney stones. While vegetable foods are alkaline, animal products are acidic, and require a huge output of hydrochloric acid from the stomach to digest them. This acid tide in the blood, after a high-protein meal, requires an equally strong base response by the body to neutralize the acid. We get the necessary alkaline contribution at the expense of our bones, which have to give up their minerals. For that purpose, our bones literally *dissolve* into phosphates and calcium.

This is a primary step in bone loss that leads to osteoporosis. Our high salt intake, then further complicates matters in contributing to the flushing of our bone mass down the toilet bowl. The excessive stimulation of bone turnover also causes an increase in bone breakdown and remodelling: that may lead to osteoarthritis and calcium deposits within our tissues.

It is physical exercise, not excess proteins, that builds muscle strength, denser bones, and bigger muscles. Of course, when you artificially stimulate growth with overfeeding and with excessive animal food, you can achieve a higher body mass index (a weight-to-height ratio), but this will add fat to your body, as well as increasing the risk of disease.

Animal Foods: High Cholesterol Levels, Triglycerides and Saturated fats

Cholesterol is essential to health, because from cholesterol the body has the ability to synthesize sex hormones and steroid hormones. Cholesterol is also important for the stability and support of cell membranes, for brain signal function (crucial role in the formation of the myelin sheath), helps towards our immune defenses and for the metabolism of fats.

Cholesterol is not the same as fat. It is a substance made in the livers of all animals. Too much cholesterol can be a dangerous substance. The human liver makes adequate amounts of cholesterol for the human body's needs. Humans *do not* need more than the body makes, and so whenever we consume foods containing high amounts of cholesterol, the body either eliminates the excess cholesterol via our stools (with the help of binding to dietary roughage or fibre, if present) or the cholesterol gets deposited into the walls of our arteries: this can lead to plaque formation, atherosclerosis and other heart-related complications.

When cholesterol is transported in the bloodstream, it is packed into special particles called chylomicrons: consisting of phospholipids (6-12%), triglycerides (85-92%), cholesterol (1-3%) and proteins (1-2%).

Chylomicrons come in many forms: low-density lipoproteins (LDL), intermediate lipoproteins (IDL), very low-density lipoproteins (VLDL) and, finally, high-density lipoproteins (HDL). They enable fats and cholesterol to move within the water-based solution of the bloodstream.

LDL is considered as "bad cholesterol" because although it is necessary in limited amounts, high LDL cholesterol levels can dramatically increase your risk of a heart attack.

HDL is considered as "good cholesterol" and can have a lowering effect to the body and so it lowers your risk of a heart attack. So, elevated concentrations of LDL predispose to heart attacks, whereas elevated amounts of HDL are considered more protective.

Dr. Dean Ornish's Explanation in the "Great Nutrition Debate"[8]

There is a genetic variability on how efficient the body can handle fat and cholesterol. The 1975 Nobel Prize winners Brown and Goldstein discovered the LDL receptor. People who have a genetic defect of Familial Hypercholesterolemia (FH), have elevated cholesterol levels in their blood many times above normal, because they have fewer LDL receptors that can help bind to cholesterol. This can increase their risk of heart attack early on in life.[9]

It is not just the fat and cholesterol in the blood that increase the risk of heart attacks, but also the quantity of LDL receptors you have. The more LDL receptors you have, the lesser risk you will have of developing heart-related diseases. This, maybe, is why some people at the end of the spectrum, which live a longer life (80-100 years) and have high LDL receptors, are at less risk of heart disease. While at the other end of the spectrum, those persons who have fewer LDL receptors, or FH, unfortunately, have a higher tendency of heart-related risks and live shorter lives.

In other words, those persons who have high LDL receptors for cholesterol can eat high-fat and high-cholesterol foods (eggs, meat and dairy products) and still may not develop atherosclerosis or heart disease. These persons are so efficient in getting rid of fat and cholesterol it sometimes does not matter what they eat. But persons who are not so efficient in getting rid of fat and cholesterol can never make it to 80-100 years. So, indeed, you do have a selective group.[10]

On the other end of the spectrum, those persons who have low LDL receptors have a higher risk of high cholesterol and heart disease, and therefore moderate changes in a diet which contain lesser amounts of dietary cholesterol does not go far enough in reducing their cholesterol levels. These persons are still eating more than their body can get rid of, and so cholesterol keeps building-up in their arteries.

It is important to note, that if a persons blood cholesterol is consistently below 150mg/dl or their ratio of total cholesterol to HDL is below 4.0 to 1, then you are either not eating enough fat or cholesterol, or your body is very good at getting rid of it. Either way, your risk of disease is so low that whatever you are doing is probably fine. Most people do not fall into this category. However, it is advisable for everyone to begin making moderate changes to their diet, until they get to this point.[11]

This is why people with low amounts of LDL receptors should not eat animal foods that contain high amounts of saturated fats and cholesterol. This is why meat, dairy and eggs are considered unhealthy foods when it comes to the risk of heart-related diseases.

The general public needs to understand that there is very little difference in comparison between the cholesterol contents of lean *white* meat and those in lean *red* meat.

Studies indicate that chicken is almost as dangerous as red meat for the heart. Chicken has about the same amount of cholesterol as beef.

Cholesterol content	Beef, top sirloin	Chicken breast, no skin
100 gm	90 mg	85 mg
100 calories	33 mg	51 mg

Regarding cholesterol, there is no advantage in eating lean white over lean red meat – as far as hypercholesterolemia is concerned.[12]

Oxidized LDL Increases Arterial-Plaque Formation that can Lead to Atherosclerosis and Health Disease

The formation of oxidative LDL occurs when LDL cholesterol particles in your body react with free radicals. Oxidized LDL itself then becomes more reactive with the surrounding tissues, producing tissue damage.

Once LDL becomes oxidized, it goes directly within the liner-lining (endothelium) of any artery in the body, including the coronary artery, carotid artery or the arteries that supply your other extremities (the arms or legs).

Once activated, oxidized LDL encourages the accumulation of inflammatory cells: such as macrophages and platelets at the vessel injured site to promote their adhesion to the damaged area. More macrophages, lipids and cholesterol begin to accumulate, which forms a plaque that grows and thickens and this can eventually lead to a blockage. These mechanisms may completely restrict blood flow to the damaged area. This long-term effect can result in a variety of health conditions, including stroke, atherosclerosis, and coronary heart disease that can lead to myocardial infarction.

What Can You Do to Prevent the Formation of Oxidized LDL?

The formation of free radicals that attack LDL is the major factor that leads to oxidized LDL, and so we obviously need to eliminate chemicals, toxins and other damaging molecules that bring about free-radical damage to our cells and tissues. There are many lifestyle and dietary changes we can do to prevent the formation of oxidized LDL, such as:

1. Stop smoking
2. Reduce foods that contain excess iron: as iron is a "two-way sword". Meaning, for many reasons, iron is essential to human health, but too much iron found in animal products or by using iron supplements can increase the build-up of free radicals and can lead to DNA damage, which can also increase oxidized LDL. Luckily, the same diet (a plant-based diet) that lowers cholesterol levels also lowers iron. Cutting meat doesn't just cut our cholesterol and fat: it also cuts out some of the iron that can damage the heart. "Meat is a one-two punch," said Harvard University biochemist Randall Lauffer. "It contains a certain form of iron that is very rapidly and easily absorbed. And it contains saturated fat and cholesterol. So every bite of meat is contributing to two problems in the body –both which may contribute to heart disease and possibly other chronic diseases that are common in Western meat-eating atherosclerosis.[13]
3. Exclude animal products which contain excess amounts of saturated fats, cholesterol, iron and other carcinogenic chemicals that increase the build-up of free radicals.
4. Add fruits, vegetables and grains to your diet. Not only do they contain plenty of nutrients, are low in fat and cholesterol and other harmful toxic substances, but more importantly, plant food sources also contain the vital components that help bind to free-oxygen species. These essential plant constituents – which I have already thoroughly mentioned, are vital to life in humans, and there are thousands of "phytochemicals and antioxidants" found in plants; they protect plants from free-radical damage, as much as when we eat plants, they also protect us.

Fish and Other Animal foods Contain Cholesterol

All fish products contain significant amounts of cholesterol, too. Shellfish, such as lobster, crayfish, and shrimp are higher in cholesterol, ounce for ounce, than beef. Eggs are packed with cholesterol. A single egg yolk contains 213 mgs. That is a huge amount and the highest concentration in any animal food.

Also, fish does contain the essential omega-3 fatty acid, but omega-3 is an unstable molecule that oxidizes easily: causing the production of free radicals. Omega-3s are needed by the body for healthy nerves, skin and the eyes, and as raw materials for building other biological molecules, but vegetables such as spinach, broccoli, lettuce, and beans provide omega-3s in a form that is more stable and also more modest in quantity.[14]

Cholesterol has been called "the animals' revenge" because every time you eat animal food they leave a little bit of them in your arteries (saturated fat and cholesterol), and they end up killing you: because their flesh was never intended for human beings.[15]

The Cholesterol and Cancer Correlation

The only source of dietary cholesterol on the planet is derived from animal foods, and there has been a very long association between high cholesterol levels and high cancer rates.[16] Studies have shown that cancer cells also have a higher demand for cholesterol than normal cells.[17]

Growing cancer cells lose their capacity to synthesize cholesterol and become dependent upon outside sources, i.e., blood cholesterol. Studies have shown that by lowering cholesterol levels, tumor growth is slowed, the spread of cancer is reduced, and survival time is prolonged.

In one study, a cholesterol-free diet cut tumour growth rates by 50 percent, and survival times were often doubled. When cholesterol was re-introduced into the patient's diet, however, the tumours started to re-emerge and grow back.[18]

But when animals were injected with human prostate tumour cells and fed high-cholesterol diets, cholesterol tended to accumulate in the outer membranes of the tumour cell. This can alter the chemical signalling patterns within the cells. As a result, human-body cells oppose signals telling them to die normally and instead continued to proliferate in an uncontrolled manner. Cholesterol did not trigger new cancers in the experimental animals. Evidently, six weeks after the tumour cells were injected, the animals on the high-cholesterol diets had twice as many tumours as animals on ordinary diets, and their tumours had grown substantially. When the cholesterol was lowered, cancer cell death increased and the tumours stopped proliferating (multiplying).

Whenever tumour cell membranes were replenished with cholesterol, the cancer ran out of control again.[19] In other words, normal cholesterol blood levels were not causing new tumours: high-cholesterol diets were "feeding" existing cancers.

Remember, "There is no hidden cholesterol in plant foods. If you do eat animal-based foods, the fibre and other essential nutrients found in plants also help to bind or flush-out excess animal cholesterol from the human body".

Triglycerides are another type of fat stored by the body. Chemically, triglycerides are made up of three molecules of fatty acids that are combined with a single molecule of the alcohol glycerol. The prefix "Tri" refers to the fact that there are three fatty acids attached to a glycerol molecule, making the compound a glyceride. Triglycerides molecules are assembled in the liver, packed in chylomicrons and VLDL, and sent via the bloodstream to the hips, thighs and abdominal fat areas, where they wait until they are needed for energy and are converted to other energy sources, like glucose.

In the intestine, triglycerides are split into monoacylglycerol and free fatty acids in a process called "lipolysis", with the secretion of lipases and bile, which are subsequently moved to absorptive enterocytes, cells lining the intestines. The triglycerides are rebuilt in the enterocytes from their fragments and packaged together with cholesterol and proteins to form "chylomicrons".

These are excreted from the cells and collected by the lymph system and transported to the large vessels near the heart before being mixed into the blood. Various tissues can capture the chylomicrons, releasing the triglycerides to be used as a source of energy.

Fat and liver cells can synthesize and store triglycerides. When the body requires fatty acids as an energy source, the hormone glucagon signals the breakdown of triglycerides by hormone-sensitive lipase to release free fatty acids.

As the brain cannot utilize fatty acids as an energy source (unless converted to a ketone), the glycerol component of triglycerides can be converted into glucose, via gluconeogenesis, for brain fuel when it is broken down. Fat cells may also be broken down for that reason – if the brain's needs ever outweigh the body's needs.

Triglycerides cannot pass through cell membranes freely. Special enzymes on the walls of blood vessels called "lipoprotein lipases" must break down triglycerides into free fatty acids and glycerol. Fatty acids can then be taken up by cells via the fatty acid transporter (FAT).

Consuming simple sugars and refined carbohydrates, such as white flour, white potatoes and white rice, can create and lead to a rapid spike in blood glucose levels. This causes the pancreas to release more insulin. When your body has more glucose than it needs for energy and has reached its storage capacity for glycogen, the increased insulin prompts the liver to convert glucose into triglycerides, which are then transported to fat cells.

Whenever a person consumes high amounts of refined sugary foods, it often means that a person is consuming less dietary fiber. This is important because increasing fibre in the diet is associated with decreasing energy intake (calories), which can result in weight loss. For those who are at a greater than ideal body weight, weight loss is an important therapy for lowering triglycerides.

If you are eating more calories than your body needs, your body may store too many triglycerides. People who consume a low-fat, high carbohydrate diet tend to have very low cholesterol levels and a lower risk of heart disease. But their triglyceride levels are elevated. This has often puzzled medical scientists. What could be happening is that the excess carbohydrates in people's diet are simply being converted to triglycerides for transport and storage in the body. This is normal, and apparently not related to heart disease. If your cholesterol is normal, an elevated triglyceride level does not appear to be a risk factor for heart disease.

Elevated triglyceride levels (greater than 1,000mg/dl), however, have been associated with diabetes and pancreatitis. Some foods raise triglyceride levels more than others, and knowing which foods cause them to elevate can help you avoid them, allowing you to keep your triglyceride levels in check.

Animal products raise triglyceride levels because they contain saturated fats and cholesterol, and so do simple sugars, like white foods, cookies, cakes, sweets and other high sugary food sources. Likewise, plant foods have the tendency to lower all known fats and triglycerides. This has become unmistakably obvious "because" plant foods contain low levels of refined sugars and harmful fats, are high in dietary fibre and antioxidants and, more importantly, contain no cholesterol. All this, overall, assists in reducing the stimulation and growth of tumor cells and does not promote heart disease.

Saturated fats are easy to recognize as they are all solid at room temperature (beef, chicken, fish, eggs, butter, margarine, etc.). Although unsaturated fat varieties are usually in a liquid form (oils), vegetable oils are naturally high in saturated fats (palm oil, kernel oil, and coconut oil). Oils can also be chemically saturated by a process called "hydrogenation". Solidified fats can then be used in products such as margarine and butter and, like animal fats; they will stimulate the liver to produce cholesterol.

Saturated fat can increase your cholesterol levels – although, some unsaturated fats have health problems of their own, including a tendency to impair the immune system and increase free radical production.[20] Liquid vegetable oils are naturally better than animal fats and tropical oils, but all fats and oils are natural mixtures of saturated and poly-unsaturated fats.

While vegetable oils do less harm than animal fats, none of them do your arteries any good. The chart below shows the percentages of saturated fat in different types of fats and oils.

Animal Fats		Vegetable Oils		Tropical Oils	
Beef tallow	50%	Cotton seed oil	26%	Coconut oil	87%
Chicken fat	30%	Peanut oil	17%	Palm kernel oil	82%
Pork fat (lard)	39%	Soybean oil	15%	Palm oil	49%
Turkey fat	30%	Sesame oil	14%		
		Corn oil	13%		
		Olive oil	13%		
		Canola oil	12%		
		Sunflower oil	10%		
		Safflower oil	9%		

Source: J.A.T. Pennington, Bowes and Church's Food Values of Portions Commonly Used (New York: Harper and Row, 1989)

The best oil to use is sunflower and safflower oil: these oils contain lower amounts of saturated fats. But if you truly want to eliminate your risk of cancer and heart disease, it is best to employ alternative cooking methods: instead of using any animal/vegetable fats and oils. It is healthier to bake, boil or steam your foods by using a steamer, a pressure cooker, a cooking bowl or a pan with water. You might lose some essential nutrients, but you will certainly not add more unhealthy fats or oils to your diet. Oil in salad dressing can be replaced with vegetable stock or water. You can also add seasoned rice vinegar or other types of vinegars to salads, instead of adding olive oil or any other plant or animal-based oil.

It is better to cook your vegetables and rice with water, and add balsamic vinegar with flaxseeds or walnuts to vegetable/nut salads: grounded flaxseeds, hemp seeds, chia seeds, and unroasted walnuts

> **FACT BOX**
>
> **All oils, whenever heated at high temperatures, are denatured into "trans-fats", and these fats are detrimental to the human body.**

and all excellent sources of omega-3 and omega-6 essential fatty acids. I always try to add nuts and grounded seeds to salads, as this, in effect, aids towards the bioavailability and transport of many essential nutrients.

The healthiest fats are the essential fatty acids (omega-3 and omega-6). These fats are vital for optimal health and should be consumed on a daily basis. We do need some monounsaturated fats, but we should all try to avoid consuming foods that contain high amounts of refined sugars, saturated fats and cholesterol. Otherwise, in the long-term, disease may prevail. By eating a variety of plant foods you will succeed in reducing your risk of heart-related diseases, diabetes, cancer and other ailments.

The Correlation between Carcinogenic Nitrosamines in Meat, Plant-Foods and Cancer

The story begins on a Norwegian fur farm back in 1957: where Mink were dropping dead from a malignant and new liver disease. The clue came when livestock too started to die from liver cancer.[21]

Throughout 1961 and 1962, in Norway there were outbreaks of toxic hepatosis in ruminants: demonstrating rather characteristic symptoms and liver lesions. Experiments indicated a possible connection between the disease and the feeding of meals made from herring that had been preserved with sodium nitrite. As the hepatotoxic properties of dimethyl-nitrosamine were established for several species, the researchers decided to investigate the toxicity of this compound to sheep, and its possible occurrence in herring meal. What actually tied all the cases together was the use of fish-meal in their diet (herring), which Norway had started to preserve with sodium nitrite.[22]

Subsequent research discovered that under certain circumstances nitrites can form nitrosamines, which directly attack deoxyribonucleic acid (DNA) and are universally considered to be the key carcinogens in cigarette smoke.[23] The addition of nitrites in meats and other foods were a matter of the gravest concern for human health nearly half a century ago. Now we know that nitrites added to processed meats can form carcinogenic N-nitroso-compounds, recognized as being among the most potent-chemical carcinogens.[24]

A study on pregnant women who eat hot dogs showed that such women run the risk of having children with brain tumors: the number two pediatric cancer.[25] Children who eat hot dogs also have around 10 times a greater chance to develop childhood leukemia: the number one pediatric cancer.[26]

Nitrites are actually preservatives added to cured meat – like luncheon meat, bacon, ham, hamburgers and hot dogs – to prevent and inhibit the germination of *Clostridium botulinum* spores. Nitrites, also provide meat its desirable cured colour by combining with heme-iron. Nitrosylmyoglobin is responsible for the red colour of raw cured meat. Cooking denatures globin which detaches from the heme, yielding a pink mononitrosylheme complex, the color of cooked cured meat.[27]

There are literally hundreds of studies linking cancer to cured-meats. In recent years processed meat consumption has been closely associated with bladder, endometrial, prostate, thyroid – and then all the way down the digestive tract to – throat, oesophageal, colon and rectal cancer.

In a 2007 study,[28] mailed questionnaires were completed by 19,732 patients with histologically confirmed cancer of the stomach, colon, rectum, pancreas, lung, breast, ovaries, prostate, testis, kidney, bladder, brain, non-Hodgkin's lymphoma or leukemia: The study also carried out 5,039 population controls between 1994 and 1997.

Measurement included in the study: information on patients' socioeconomic status, lifestyle habits, and diet. A 69-item food frequency questionnaire provided data on eating habits 2 years before the study. Odds ratios and 95% confidence intervals were derived through unconditional logistic regression. Compared with "never adding salt-at-table", "always or often adding salt-at-table" was associated with an increased risk of stomach, lung, testicular and bladder cancer.

Processed meat was significantly related to the risk of stomach, colon, rectum, pancreas, lung, prostate, testis, kidney, and bladder cancers and to leukemia. The odds ratios for the highest quartile ranged from 1.3 to 1.7. These findings add to the evidence that high consumption of salt and processed meat play a role in the aetiology of several cancers.[29]

According to the "Salt, Processed Meat and the Risk of Cancer" study, "processed meat was significantly related to the risk of the stomach, colon, rectum, pancreas, lung, prostate, breast testis, kidney, and bladder and leukaemia as well".[30]

The Dangers of Nitrosamines in Cured Meats

With the concern over the potential danger of nitrosamines growing, consumer groups such as the "Centre for Science in the Public Interest", actually, petitioned the United States Department of Agriculture (USDA), as far back as 1972, to ban or at least greatly reduce the nitrite in cured meats. The USDA denied the petition citing the nitrites' role in the prevention of botulism bacteria that can grow in vacuum-packed meats. They had to weigh the risk of cancer with the risk of consumers getting deadly food-poisoning bacteria from lunch meat. So take your pick: Cancer or Botulism. But there is a third option. Choose the latter, which is a "plant-based diet", without the risk of both.

Are Nitrites Pollutants or Nutrients?

Phytonutrients such as vitamin C prevent the formation of nitrosamines from nitrites. This explains why adding nitrite preservatives to processed meat can be harmful, but adding more vegetables and their nitrite-forming nitrates in our diet can be helpful.

We now know that cured meat increases cancer risk, but a high intake of vegetables is strongly associated with a reduced risk. Why? How can nitrites be bad in meat products, but good when they originate in our own mouths from the nitrites we get from green-leafy vegetables?

Let us go back to the basics. The fact of the case is that it is the nitrosamines that are the carcinogens and not the nitrites themselves. The nitrosamines are the compounding enemies to the development of cancers and other diseases. The only reason we are concerned about the nitrites is that *under certain circumstances* they can turn into nitrosamines and other N-Nitrosamines carcinogenic compounds.[31]

The nitrites themselves are fine, in fact, they are amazing! That is why most green-leafy vegetables, especially arugula (rocket lettuce) and beet juice – that contain the highest amounts of healthy nitrites, actually aid in boosting Nitric Oxide activity and reduce blood pressure. Plant-foods also reduces our risk of circulatory vascular problems – like strokes, atherosclerosis and heart disease.

The nitrates turn into nitrites, which then turn into Nitric Oxide (NO) which helps open up arteries and also boosts athletic performance.[32] So, as long as nitrites turn into NO our health will be fine. It is only when they turn into nitrosamines that they cause human health problems.[33]

So the answer to the riddle – finally – lies in the circumstances in which nitrites form nitrosamines. And that circumstance is in the absence of *plants*.[34]

Phytonutrients, like caffeic acid, ferulic acid, ascorbic acid (Vitamin C), and others, found in plant foods, aid in blocking nitrosamine formation. Nitrites together with plant foods are not a problem. But are there any of these protective compounds or vitamin C in meat? The answer is "No!"

So, if you add nitrites to meat, nitrosamines will pre-form in the meat before it even makes it into our mouths. It is not so much that we are eating the nitrites added to the meat, but eating the nitrosamines that form in the meat when they add nitrites to it. A simplified explanation is that:

→ Nitrates → Nitrites → Plants → Nitric Oxide → Opens up Arteries

→ Nitrates → Nitrites → Absence of plants → Carcinogenic Nitrosamines

"Nitrites in the absence of plants turn into carcinogenic nitrosamines."[35]

Animal Protein Increases Insulin-Like-Growth Factor-1 (IGF-1)

Why do centenarians escape cancer? Why do some people live up to a hundred years? As we get older, our risk of getting or dying from cancer increases year-by-year. But once you get to the age of 85-90, strangely enough, your cancer risks start to drop. I suppose that kind of makes sense, because if you have not got cancer by that age, then you most probably will never get it.[36]

We could say that some individuals do live a longer life, even though they do not eat healthily. But these are only a few exceptions. Some people have a greater resilience to diseases (psychological resilience or immune resilience) and better genes, so we need to take into account the whole population's statistical resilience values: but most people are afflicted more or less by the same disease.

But, if you live for that long, it can be a sign that there is something special about you. Therefore, it seems that centenarians are endowed with a peculiar resistance to cancer. So, what is their secret?[37]

Every day, 50 billion of our cells die, and everyday 50 billion new ones are born. There needs to be a balance, otherwise our body will atrophy, shrink, or get too big or crowded. Sometimes we need to grow: e.g. when you are a baby, or during puberty and adolescence. Our cells do not get larger when we grow up, but they get more numerous.

Once we grow up we *do not* want extra cells in our bodies, but we still need our cells to grow and divide: out with the old and in with the new. We just do not want to make more cells that we do not need.

When you are a child, extra growth is good, but when you are an adult extra growth can mean a tumour. How do our cells in our bodies know how to balance these essential biological, metabolic, growth factors to safeguard ourselves from normal development? One key signal is a growth hormone called Insulin-Like-Growth-Factor-1 (IGF-1).[38]

IGF-1 levels go up when you are a child and levels decrease or balance out when you are fully developed into adulthood. But if levels are kept constantly high when you are an adult, then there is a constant message to our cells to keep growing: "Grow! Grow! Grow!"

Cow's milk increases the production of IGF-1 in humans, which may lead to increased cancer growth. Tissues like the breast and the prostate gland are indeed both hormone growth-dependent. It is now believed that high amounts of animal proteins also increase the liver's production of IGF-1.[39]

Not surprisingly, the more IGF-1 in our bloodstream, the higher our risk is in the formation and development of cancer. More IGF-1 in our bloodstream, the more prostate cancer, breast cancer, etc.

But it is **not** the original tumour that tends to kill you, but it is usually whenever any malignant tumour metastasizes. IGF-1 is a potential and powerful growth factor, and it makes things grow, including cancer.

This growth factor helps cancer cells break-off from the main tumour, migrate in the surrounding tissues, and invade the bloodstream. What helps breast cancer metastasis to the bone? The answer is: IGF-1.

IGF-1 helps transform normal cells and cancer cells in the first place. The IGF-1 helps cancer: towards their survival, proliferation, transformation, self-renewal, metabolism, angiogenesis (formation of new blood vessels) and spread (metastasis). Therefore, too much IGF-1 during adulthood increases the risk of cancer.

You may have heard of a rare genetic defect that leads to a type of dwarfism (Laron syndrome), that is IGF-1 deficiency, where the growth factor levels are so low that such persons only grow to a small height – but they almost never develop cancer.[40]

Laron patients' mutation means that their growth-hormone receptor lacks the last eight units of its exterior region, so it cannot react to the growth hormone. In normal children, the growth hormone

ANIMAL PRODUCTS: THE SLOW-GROWING, FLESHY ROAD TO CHRONIC DISEASES

makes the cells of the liver churn out IGF-1, and this hormone makes children grow. If Laron patients are given doses of IGF-1 before puberty, they can grow to fairly normal height. When they looked at the IGF-1 levels in Laron's non-affected relatives (compared to those with the gene mutation and then at their relative cancer rates), researchers noticed that about 20% of their relatives died of cancer, which is pretty standard. And what was the percentage of those with IGF-1 deficiency dying of cancer? It was zero. Not a single person in the study died of cancer.[41]

Here is what non-affected family members died of: the usual heart disease, cancer and stroke. The folks with IGF-1 deficiency may die of a lot of things. They actually get hit by cars more; get tripped down the stairs, etc. So, 20% of deaths were due to accidents, but they never died of cancer.[42]

So scientists are now beginning to ask whether or not it is possible to have the best of both worlds? In retaining all the growth factors you need as a child (so you grow up to a normal height), then, as soon as you attain a normal adult stature, try to keep your IGF-1 level low (thus, keeping your cell life and death balance sheets balanced).

The truth is that we can achieve both normal stature and our cell life/death sheet balanced through dietary manipulation.[43] This can be maintained by putting people on a low-fat, high-fibre, plant-based diet with plenty of exercise: that will lower circulating IGF-1 levels, reverse and reduce the risk of cancer growth.[44]

We now know that animal protein is more associated with an increased level of IGF-1, but not plant protein. With the exception of the animal protein gelatine, all proteins, plant and animal, have all the essential amino acids and so pretty much all proteins, in that sense, are complete proteins. But when medical scientists talk about high vs. low quality proteins, what they are really referring to is the relative proportions of the different essential amino acids. The more closely the proportions match our own proteins, the higher is the quality considered to be.

When our liver gets stimulated by a large amount of animal protein, it somehow finds it easier to release IGF-1 into the bloodstream: making cells divide more aggressively and excessively. But when you ingest plant protein, the liver somehow finds the time to catch up, and it releases IGF-1 into the bloodstream more constantly, which indeed gives our cells and tissues the time to develop normally.

Basically, all the essential amino acid can be obtained in the body from the foods we eat. The body can break them down to synthesise them, to help build essential proteins, enzymes, hormones, carrier proteins (albumin). But plant proteins just do not stimulate the body to release IGF-1 like animal proteins do. It may or may not be true, but it is believed that animal protein stimulates the production and release of

IGF-1 from the liver more frequently and more aggressively, than plant-based proteins.

In the China Study, Campbell found out that animal protein increases the activity of an important enzyme in the human liver named "Mixed Function Oxidase." This enzyme can convert a normal healthy human cell into a cancer cell. Campbell concluded that "The more animal protein, the more cancer (…) We need protein, but not more than 5-10% of our diet, and specifically not coming from animal protein."

Animal proteins also increase other metabolic functions, including the impairment of the immune system, especially the important protective "natural killer cells of the immune system". Animal protein increases oestrogen levels, hormone levels, and IGF-1 and might in turn be associated with breast cancer. There are hundreds of thousands of healthy chemicals in foods that work in synergy and the final effect is either "health or disease": depending on whether you are eating animal foods or plant foods.[45]

Another important reason why animal proteins may increase our risk of developing cancer, is maybe because it has to do with digestive/pancreatic enzymes or a shortage of them.[46]

So, as the evidence clearly shows, in order for us to reduce IGF-1 and high-protein levels, we need to eat more plant-based foods and less of the other cancer-promoting agents found in animal products.

Breast Cancer Risk by Heterocyclic Amines (HCAs) and Polycyclic Aromatic Hydrocarbons (PAHs)

Heterocyclic amines are carcinogens in cooked meats. Seventy years ago a Swedish researcher first reported that feeding mice with roasted horse muscles caused cancer in the mice. This cancer-producing substance has since been identified.[47]

Heterocyclic amines are the carcinogenic chemicals formed during the cooking of meats: such as beef, pork, poultry, fish, etc. They are mainly created when the building blocks of muscle tissue, amino acids (the building blocks of proteins), sugars, and creatine (a substance found in muscles) react at high temperatures: during roasting, frying, grilling and barbequing.[48]

Thankfully, plant products tend *not* to produce these carcinogenic heterocyclic amines.

Seventeen different such carcinogens have so far been identified and discovered in cooked meats. The National Cancer Institute goes on to explain how people eating "well-done" meat appear to have a higher cancer risk, than those eating meat cooked "rare".

But remember that we are not meant to eat "rare" meat anymore, due to the risk of food poisoning and "now", evidently due to the risk of cancer and atherosclerosis, etc.

The reason we are so concerned these days about these cooked meat carcinogens is that – in 2007 it was discovered – that humans are much more susceptible to these harmful compounds than we may

have thought. Previous research was conducted on rats. Rodents have this uncanny ability to detoxify 99% of the heterocyclic amines the researchers feed them. But researchers have discovered that the human liver can only detoxify 50% of the carcinogens they get from cooked chicken.[49]

This means that instead of 1% getting into our blood system (as in the case of rats), medical scientists now know that around 50% actually gets absorbed into the human bloodstream. So now we all should be 50 times more aware and concerned about our health!

As we have said above, heterocyclic aromatic amines are carcinogen compounds naturally formed during cooking of meat. The abundance of antioxidants found in plant foods can be a useful protective strategy in reducing the carcinogenic effects of these compounds. So, whenever you eat meat you should *always* combine healthy-plant foods with your meal, which will counteract the effects of the build-up of heterocyclic amines in cooked meat.

So, it is advised to lessen, or better completely avoid, your daily intake of cooked meat products, owing to the buildup of heterocyclic amines that are found therein. Heterocyclic amines have been proven to be highly carcinogenic and a contributing factor to an increased cancer risk and other pathological diseases.

But if we know that *heterocyclic amines* (HCAs) are carcinogens formed in or on the surface of well-done meat, cooked at high temperature; then what about the other carcinogenic *polycyclic aromatic hydrocarbons* (PAHs)?[50]

PAHs are compounds found in cigarette smoke, and are also formed when muscle meat (including beef, pork, fish, or poultry) are cooked using high-temperature methods, such as pan frying or grilling directly over an open flame.[51] In laboratory experiments, HCAs and PAHs have been found to be mutagenic, that is, they cause changes in DNA that may increase the risk of cancers, mutations and birth defects.[52]

In a 2007 study, researchers studied breast cancer risk in relation to intake of cooked meat in a population-based, case-control study

(1,508 cases and 1,556 controls) conducted in Long Island, NY from 1996 to 1997. Lifetime intakes of grilled or barbecued and smoked meats were derived from the interviewer-administered questionnaire data. Their results supported the already accumulating evidence, that consumption of meats cooked by methods that promote carcinogen formation, may increase the risk of postmenopausal breast cancer.[53]

But, why more breast cancer risk? These cooked meat carcinogens are mutagenic, meaning they damage human DNA. In fact, you can directly correlate the number of DNA mutations in human-breast tissue with estimates of dietary intake. The researchers asked women undergoing breast reduction surgery about their meat-cooking methods. They concluded that "consumption of processed meat – as well as fried, stir-fried meat, and heterocyclic amines (HCA) intake, were correlated with the level of DNA mutations they found subsequently in the women's breast tissue".

But we already know that these HCA's can damage DNA. After being absorbed into the body through the skin or lungs, PAH compounds are distributed to almost all internal organs and are also transferred into the placenta. It was shown that newborns and young children are especially vulnerable to the toxic effects of airborne PAH. It has been recently documented that prenatal exposure to airborne PAH may also effect the future cognitive development of children. We should *never* smoke around children. Perhaps we should also stop grilling meat around them too!

Even just living next-door to a restaurant preparing meat can pose a hazard. In a 2012 study,[54] 20 restaurants situated in mixed residential and commercial communities were selected. These restaurants were categorized as Chinese (9 restaurants), Western (7 restaurants), and Barbeque (4 restaurants). None of the restaurants allowed smoking indoors. The fumes from cooking in all the restaurants were extracted through an exhaust hood and discharged into the atmosphere near the back neighbours. So they compared what was coming out of Chinese, American and BBQ establishments. Which restaurants do you think were the worst in release PAH into their neighbors atmosphere?

The worst were the Chinese restaurants! It may come as a surprise that it was not the BBQ places. The researchers think it was because of the fish. More fish are broiled in Chinese restaurants than in the BBQ establishments. They calculated that given the excess cancer risk you would not like to be living next to a Chinese restaurant. They concluded that it may be more risky for adults, since they inhale more, ingest more, and have more skin exposed to carcinogens than children. Instead of trying to breathe less, though, it might be better to move or convince the restaurant to "go vegetarian"!

In a 2004 study, what really surprised everyone was that not only these meat chemicals could be triggering the original cancer-causing mutation, but they may promote the growth of the sewing tumour, as 2-amino-1-methyl-6-phenlimidazo [4,5-b] pyridine (PhIP) was discovered to be a potent oestrogen.[55]

When they dripped the kind of level of PhIP you would expect in your body after eating cooked meat, the researchers found that PhIP actually activated oestrogen receptors almost as powerfully as straight normal human oestrogen. That is what they found when they tried it on breast cancer cells.

They found the PhIP proliferative potency on human breast cancer cells approaching that of pure oestrogen. They concluded that:

> PhIP possesses oestrogenic activity at low concentrations, supporting the idea that exposure to PhIP, even at low doses, could have oestrogenic effects. They suggested that the well-established and unequivocable genetic toxicology of PhIP, combined with its oestrogen activity, could drive clonal expansion and promote growth of the initial cancer cell.[56]

But these were breast cancer cells in a Petri dish. How do we really know that when we consume meat these carcinogens can actually travel and get deposited not only in breast tissue, but also in the breast ducts where most breast cancers arise (ductal carcinoma)?

Nobody knew for sure until this 2001 study was performed.[57] The study measured the levels of PhIP in breast milk formed in the ducts of non-smoking women. The average concentrations they found in the breast milk of meat-eating women, corresponded to significant breast cancer growth potential. One of the women in the study actually was vegetarian, and the researchers concluded that they could not find any concentration of PhIP present in her breast milk.

This finally concludes that persons who eat cooked meat retain more HCA's and PhIP (like oestrogen) in their blood and within breast tissue. These potent carcinogens may then indeed increase the risk towards the development of breast cancer and other cancers, especially in cooked meat.

PhIP: The Three Strikes of Breast Carcinogen

As we have learned, the cooked meat carcinogen PhIP:

- Causes DNA mutations that may initiate a tumor;[58]
- Promotes the growth of the cancer due to its potent estrogen activity;[59]
- May then promote the invasiveness of breast cancer cells (metastasis).[60]

The most frequent cause of death in patients with breast cancer is due to the metastasis of cells from the primary tumour to distant sites: such as the liver and the lung, where they can grow to form secondary tumours.[61]

The way you can test for cancer invasiveness is by placing cancer cells in what is called an "invasion chamber".

Cancer cells were placed, to the top on one-side of a membrane containing tiny pores. In the study on the underline of the filter no invasion was detected, but when some oestrogen was added the researchers noticed cancer cells proliferating and growing within the chamber. However, when they added some PhIP to the chamber, there occurred an even greater breast cancer cell invasion.

From this, the researchers concluded in the study that:

> This present study and our previous reports have demonstrated that in addition to PhIP well characterized genotoxic potential (DNA mutation-

causing potential) PhIP is potently estrogenic, capable of powerful hormonal activity, and is able to potently stimulate breast cancer cells to invade through a basement membrane model. This finding, that PhIP is able to exert this pro-invasive phenotype in breast cancer cells at such low concentrations, is remarkable. The genetic toxicology of the compound to its ability to enhance cell proliferation and invasion indicates that PhIP can act not only to initiate the carcinogenic process but also to promote it.[62]

The researchers added that the exposure to PhIP is difficult to avoid because of its presence in many commonly consumed cooked meats: particularly chicken, beef and fish.[63] But PhIP has also been identified in cigarette smoke, diesel exhaust particles, and incineration ash, suggesting that it may also be a widespread-environmental pollutant.

The cooked meat carcinogens, implicated in promoting the initiation, growth and spread of breast cancer, may also increase the risk of prostate cancer. The mechanism through which the consumption of "well done" meat may increase prostate cancer risk is via the release of mutagenic compounds during cooking. Just to remind you all again. Heterocyclic amines (HCAs) and polycyclic aromatic hydrocarbons (PAH) are chemicals formed when muscle meat such as beef, pork, fish or chicken (birds) are cooked at high temperature methods, such as pan frying or cooking over an open flame (barbequing).[64]

In the case of chicken, it has been proven that the temperature during cooking does not have to be so high. Even *baking* at 180°C for 15 minutes can lead to significant production of HCAs, including PhIP. This means that harm is not done only by frying, barbequing or grilling.

These cooked meat carcinogens have recently been associated also with an increased risk of kidney cancer, colon cancer, lung cancer, breast cancer and pancreatic cancer. Pancreatic cancer is not a cancer men want to develop, because it is one of the hardest cancers to cure. In most cases, pancreatic cancer is rapidly fatal.[65]

How do we decrease our exposure to these potent mutagens? Fried bacon and fish are the worst. Strangely enough, chicken without the skin has higher mutagen levels than chicken with skin. Researchers opine that "the skin present at the surface acts like an insulating layer of the meat".[66]

"Medium-to-rare" meat is less mutagenic than "well done" meats. This could be why women in the Iowa health study; consumed meat "very well done" appeared to have a 4.62 times higher risk, than women who consumed their meats "rare" or "medium" done.

But this raises the so-called paradox of preparing meat, noted by the "Harvard Health Letter" that runs like this: well-cook meat, and you risk cancer; under-cook, and risk *Escherichia coli* food poisoning. This means that eating boiled meat is probably the best and the safest way to eat meat. If you eat meat that never goes above 212° Fahrenheit (100°C) then both your urine and your faeces are significantly less mutagenic, compared to eating meat cooked at higher temperatures. This means that you have less DNA-damaging substances circulating through your bloodstream and coming in contact with your colon.[67]

If you want to seriously minimize your exposure to PhIP it is best not to overcook your meat. The better solution would be to reduce your meat intake to once or twice a month only. And if you want to completely eliminate heterocyclic aromatic amines and PhIP, it is advised to eat solely, "plant-based foods."

Tamoxifen and Insulin-Like Growth-Factor-1 (IGF-1): A Ticking-Time Bomb!

AstraZeneca is a global innovation-driven biopharmaceutical company: specialising in the discovery, development, manufacturing and marketing of prescription medicines. The company says that it actually makes a meaningful difference in healthcare. AstraZeneca manufactures the controversial, and widely prescribed, breast cancer drug "Tamoxifen".

In 1992, the American Cancer Society launched an aggressive marketing programme with the National Cancer Institute, designed to introduce the public to the concept of "chemo-prevention" – the highly dangerous practice of taking chemotherapy drugs to prevent cancer in supposed "high risk females" whom, at the time were considered as cancer-free.

This programme aimed at recruiting 16,000 females in a five-year trial using the highly profitable Zeneca cancer drug Tamoxifen. The women were told the drug was essentially harmless and that it could reduce their risk of contracting breast cancer. Yet, these women were not told that Tamoxifen, described by Gary Williams, Director of the American Heart Foundation, as "a rip-roaring liver carcinogen." Tamoxifen has been demonstrated to be a potent cancer-causing agent in rodents and able to induce human uterine cancers.[68]

In his book, "Indicated Cancer Research", Dr. Tibor J. Hegedus writes that:

> Tamoxifen is given to women with breast cancer to block the entrance of estradoil into the tumour cells dependent upon this hormone to stimulate growth. When these hormones are blocked from reaching their primary targets, they are forced to travel to other organs. This, in turn, stimulates proliferation of cells in the lining of the womb and may cause the increased risk of endometrial cancer.[69]

Tamoxifen is widely used today in the treatment of breast cancer. In May 1995 the drug was added to California's Carcinogen Identification list in compliance with Proposition 65, a law that requires the state to publish and maintain a list of all known carcinogens. Tamoxifen's dark side had been the subject of many an investigational piece.

Several studies, including one reported to the FDA's Oncological Drugs Advisory Committee by the National Surgical Adjuvant Breast and Bowel Project in 1991, showed that the risk of developing life-threatening blood clots increases about sevenfold in women taking Tamoxifen.[70]

In April 1996, the World Health Organization (WHO) actually declared Tamoxifen to be a carcinogen, but AstraZeneca continues to market this toxic drug. In May 2000, the New York Times reported that the National Institute for Environmental Health Sciences listed substances that are known to be carcinogenic, and Tamoxifen was included in that list. Taking a carcinogen to try to stop the spread of cancer is like giving cigarettes to a patient who has already been diagnosed with lung cancer.

Robert Cohen, in his book "Milk: The Deadly Poison", states:

"The single most disturbing aspect of rBGH, from a human safety standpoint, concerns IGF-1, which is linked to breast cancer."[71]

And also according to Dr. Samuel Epstein:

"IGF is not destroyed by pasteurization, survives the digestive process, is absorbed into the blood, and produces potent growth promoting effects."[72] Epstein says it is very likely that IGF helps transform normal breast tissue to cancerous cells, and enables malignant human breast cancer cells to invade and spread to distant organs.[73]

- Breast cancer cell cultures respond to minute IGF-1 concentrations by multiplying as much as 4-5 fold.[74]
- Nearly all breast cancer cell lines (cultures) and breast cancer cells from fresh tumour biopsies have receptors for IGF-1, and binding of IGF-1 to breast tumours is increased compared to normal breast tissues. Median IGF-1 concentrations in primary breast cancers have been found to be significantly higher, than in normal breast tissue.[75]
- Over-expression of IGF-1 receptors is suggested to be the key factor in transforming normal breast tissue to breast cancer. IGF-1 also causes changes in the cell cycle and cancer-causing genes, "oncogenes". Minute concentrations of IGF-1 alter the relative number of breast cancer cells in each phase of the breast cancer cell cycle, and such alterations may indeed cause the unregulated growth of cancer. Finally IGF-1 may also have the ability to make transformed cells more responsive to signals from other growth factors.[76]
- Humans manufacture their own IGF-1, but could disease be caused by consuming extra quantities of IGF-1 from dairy produce and the meat of animals used in dairying? According to Macaulay (1992), although IGF-1 is a normal body component, it may well be associated with malignant disease when present in excess.[77]
- IGF-1 had growth-promoting effects in response to concentrations as low as 1ng/ml. Milk contains approximately 30ng/ml so that two, 8oz glasses of milk a day would contain 200ng of IGF-1 per kilogram

of body weight per day for a person weighting 70kg.[78] Also, milk secretions of mammals contain specific forms of IGF-1 that is ten times more potent than normal IGF-1. In normal cow's milk 3 percent of the IGF-1 is reported to be in this form.[79]

- Studies show, however, that growth hormones contained in milk closely similar to IGF-1 are not destroyed by the GIT because of the protective effect of casein (the principal protein in milk).[80] IGF-1 and other growth components of cow's milk would therefore survive digestion and risk being absorbed into the human bloodstream. Once in the bloodstream, such chemicals would be able to affect breast and prostate tissue, and could also promote cancer cells wherever they have reached or metastasized.

It is little wonder that dairy products have been strongly linked to breast and prostate cancer and other hormone-dependent cancers. Again, excess body fat can also stimulate the production of estrogens, so the more body fat you have, the higher your estrogen levels. This may be why overweight women have the tendency of having a higher risk of developing breast cancer.[81]

Combining Tamoxifen and IGF is like placing a ticking time-bomb underneath a table while someone is just about to have his dinner. Do you get the picture? Can you imagine anyone using both rBGH in cow's milk and Tamoxifen? You would be surely fertilizing breast cancer tumour cells to grow and reproduce, and which may evidently spread to other organ sites.[82]

Mixing the usage of Tamoxifen, daily products and animal fat is just like slowly waiting for a ticking time bomb to detonate, that may lead to breast, prostate and other cancers.

Conclusion

Medical scientific literature has provided us the evidence about the ill benefits of animal products and their contribution towards human disease.

If a person is trying to rebuild a healthy body, but is still eating animal-based foods, then these foods may not only promote cancer, but cardiovascular and a host of other diseases.[83] If you are trying to get the biochemistry of the body back into balance, consuming animal foods will most certainly unbalance it. This is because most animal-based foods all contain high levels of:

- Saturated Fats
- Cholesterol
- Protein
- Sodium
- Iron (excess iron can bring about toxic overload and oxidative stress)
- Growth hormones and in some cases also antibiotics
- Pesticide residues and other harmful toxic chemicals
- Arachidonic acids (turned into prostaglandins, linked to cancer)
- Lead, cadmium, mercury and other metals (mainly found in sea foods)

We must also not forget that animal foods contain more pesticide residues, can increase the production of IGF-1 (which is a cancer promoter), and bring about heterocyclic amines, whenever meat is cooked at high temperatures. Evidently, whenever sodium nitrites are added to certain animal foods as preservatives, in the absence of plant foods, nitrates turn into nitrites, which can evidently build-up and transform in carcinogenic nitrosamines.

These are some of the many reasons why animal-based foods are harmful to human health, and these health issues need to be addressed.

The only thing you'll miss from not eating animal foods is the saturated fat, cholesterol and animal proteins – as well as the diseases these ingredients afflict to the human body. Three of the major nutritional weapons in your personal war against cancer are antioxidants, phytochemicals, including dietary fibre. Where do these come from? Plant foods! Let us help the body fight back against disease and not increase its risk. It's just common sense. If you eat "animal products" your body will eventually suffer the consequences that leads to a slow-growing, fleshy-road to chronic disease.

Any diet which recommends eating animal foods to fight cancer is recommending a "sub-optimal" diet, to be kind, and a just plain dangerous diet, to be unkind.[84]

Removing animal products from your daily dietary regime and opting for plant-based foods, has been scientifically documented in thousands of studies, as the best solution to improve health and in reducing the risk of many degenerative diseases, especially cancer.[85] Health is the greatest gift of life, without it; there can never be full gratification or happiness.

Prevent disease – plant foods or promote disease – animal foods. It is now up to you to choose.

Notes for chapter 25

1. Campbell, T.C., & Campbell, T.M. (2006). The China Study: Part 1: The China Study: Turning off Cancer, pp.48-56. Publihsed by BenBella Books.
2. Preston, K. *A Cure for Cancer? Eating A Plant-Based Diet.* http://www.huffingtonpost.com/kathy-freston/a-cure-for-cancer-eating_b_298282.html
3. Campbell, "The China Study", 2006.
4. Campbell, T.C., & Campbell, T.M. (2005). The China Study: Part 1: The China Study: Turning off Cancer, pp.46-65. Publihsed by BenBella Books.
5. Ibid.
6. Ibid.
7. Barnard, N. (May 1994). *Food for Life. How the New Four Food Groups Can Save Your Life.* Chapter 6: The New Four Food Groups and How They Work. pp.142-146. Published by Harmony.
8. Ornish, D. (February 24th, 2000). *The Great Nutrition Debate.* "The Great Nutrition Debate" that was sponsored by the USDA. The debate features speakers (from left to right) such as Dr. John McDougall, Dr. Robert Atkins, Dr. Barry Sears, Dr. Denise E. Bruner,

Dr. Dean Ornish, Dr. Morrison, C. Bethea, Dr. Keith T. Ayoob, and a small presentation by Dr. Eric Westman. Retreived from http://www.youtube.com/watch?v=J29eNPohRyk

9   Goldstein, J.L. & Brown, M.S. (2009). History of Discovery : The LDL Receptor. *Arteriosclerosis, Thrombosis, and Vascular Biology*, **29**, pp. 431-438.

10 Barnard, N., ibid.; Ornish, D., ibid.

11 Ornish, D., ibid.

12 Davidson, M.H *et al.* (1999). Comparison of the effects of lean red meat vs. lean white meat on serum lipid levels among free-living persons with hypercholesterolemia: a long-term, randomized clinical trial. *Arch Intern Med*, **159**(12), pp.1331-1338.

13 Barnard N. (May 17th, 1994). Food for life: Chapter 2: Preventing and reversing heart disease, pp. 42-44. Published by Harmony, Reprint Edition

14 Hunter JE. (May, 1990). n-3 fatty acids from veteable oils. *Am J Clin Nutr*, **51**(5), pp.809-814. Renaud, S *et al.* (1986). Influence of long-term diet modification on platelet function and composition in Moselle farmers. *Am J Clin Nutr*, **43**, pp.136-50.

15 William Roberts, M.D (editor of The American Journal of cardiology) Retreived from http://www.all-creatures.org/quotes/roberts_william.html

16 Ibid.

17 Sidney, S. (September, 1983). Cholesterol, Cancer, and public health policy. *The American Journal of Medicine*, **75**(3), pp.494-508. Cruse, J.P *et al.* (July, 1982). Dietary cholesterol deprivation improves survival and reduces incidence of metastatic colon cancer in dimethylhydrazine-pretreated rats. *Gut*, **23**(7), pp.594-599. doi:10.1136/gut.23.7.594. Chen, H.W. (1978). The role of cholesterol in malignancy. *Pro Exp Tumour Research*, **22**, pp.275-316.

18 Mady, E.A. (December 31st, 2000) Association between estradiol, oestrogen receptors, total lipids, triglycerides, and cholesterol in patients with benign and malignant breast tumours. *J Steroid Biochemistry Molecular Biology*, **75**(4-5), pp.323-328.

19 Zhuang, L *et al.* (2005). Cholesterol targeting alters lipid raft composition and cell survival in prostate cancer cells and xenografts. *Journal of Clinical Investigations*, **115**(4), pp.959-968.

20 Zhuang, "Cholesterol targeting alters lipid raft composition and cell survival", 2005.

21 Koppang, N *et al.*(1957). A Survey of Feeding N-Nitrosodimethylamine (NDMA) to Domestic Animals over an 18 year period. *National Veterinary Institute, Oslo*. Koppang, N. (1987). Vascular changes and liver tumours induced in mink by high levels of nitrite in feed. *National Veterinary Institute, Oslo*. IARC Scientific Publications: 84, pp.256-260. Sakshaug, J *et al.* (June 19th, 1965). Dimethylnitrosamine; its hepatotoxic effect in sheep

and its occurrence in toxic batches of herring meal. *Nature*, **206**(990), pp.1261-1262. Greger, M. (February 24th, 2012). *When Nitrites go bad*. Volume 7 video. Retreived from http://nutritionfacts.org/video/when-nitrites-go-bad/

22 Greger, "*When Nitrites go bad*", 2012.
23 Ibid.
24 The Lancet. (May 18th, 1968). Nitrites, Nitrosamines, and Cancer, **291**(7551), pp.1071-1075. doi:10.1016/S0140-6736(68)91418-9. Pegg, R.B. (2004). *Nitrite Curing in Meat. The N-Nitrosamine problem and Nitrite Alternatives* (1st ed). Published by Wiley-Blackwell.
25 Huncharek, M et al. (2004). A Meta-Analysis of Maternal Cured Meat Consumption during Pregnancy and the Risk of Childhood Brain Tumours. *Neuro Epidemiology*, **23**(1-2): pp.78-84.
26 Peters, J.M et al. (March, 1994). Processed Meats and Risk of Childhood Leukaemia (California, USA). *Cancer, Causes and Control*, **5**(2), pp.195-202.
27 Ibid.
28 Jinfu Hu, et al. (March, 2011). Salt, processed meat and the risk of cancer. *The European Journal of Cancer prevention*, **20**(2), pp.132-139. This study assessed the association between salt added at table, processed meat, and the risk of various cancers. doi: 10.1097/CEJ.0b013e3283429e32.
29 Ibid.
30 Ibid.
31 Greger, M. (February 28th, 2012). *Are Nitrites Pollutants or Nutrients?* Volume 7 video. Retreived from http://nutritionfacts.org/video/are-nitrates-pollutants-or-nutrients/
32 Franzin, L et al. (January 2012). Food selection based on high total antioxidant capacity improves endothelial function in a low cardiovascular risk population. *Nutrition, Metabolism, and Cardiovascular Diseases*, **22**(1), pp.50-57.
33 Greger, "*Are Nitrites Pollutants or Nutrients?*", 2012.
34 Bartsch, H et al. (1998). Inhibitors of endogenous nitrosation: Mechanisms and implications in human cancer prevention. *Mutation Research*, **202**, pp.307-324.
35 Greger, "*Are Nitrites Pollutants or Nutrients?*", 2012.
36 Greger, M. (September 25th 2012). *IGF-1 as One-Stop Cancer Shop*. Volume 10 video. Retreived from http://nutritionfacts.org/video/igf-1-as-one-stop-cancer-shop/
37 Salvioli, S et al. (2009). Why do centenarians escape or postpone cancer? The role of IGF-1. *Cancer Immunol*, **58**, pp.1909-1917.
38 Greger, "*IGF-1 as One-Stop Cancer Shop*", 2012.

39   Yu H. (September 20th 2000). Role of the Insulin-like Growth Factor family in cancer development and progression. *Journal of the National Cancer Institute*, **92**(18): pp.1472-1489. Holmes, M.D. (2002). Dietary Correlates of Plasma Insulin-like Growth Factor 1 and Insulin like Growth Factor Binding Protein 3 Concentrations. Cancers Epidemiol Biomarkers Prev, **11**(9): pp.852-861.

40   Wade, N. (2011). *Ecuadorean Villagers May Hold Secret to Longevity. Science*, The New York Times. Retreived from http://www.nytimes.com/2011/02/17/science/17longevity.html?_r=0

41   Guevara, A *et al*. (February 16th 2011). Growth hormone receptor deficiency is associated with a major reduction in pro-aging signaling, cancer, and diabetes in humans. *Science Translational Medicine*, **3**(70), p.70.

42   Ibid.

43   Greger, M. (September 26th 2012). *Cancer-Proofing Mutation*. Volume 10 Video. Retrieved from http://nutritionfacts.org/video/cancer-proofing-mutation/

44   Barnard, R.J *et al*. (2006). Effects of a low-fat, High-Fibre Diet and Exercise Program on Breast Cancer Risk Factors In Vivo and Tumour Cell Growth and Apoptosis In Vitro. *Nutrition and Cancer*, **55**(1). pp.28-34.

45   Campbell, T.C. (2008). *Healing cancer from inside out*. DVD 2 Video

46   Lipinski, B *et al*. (2000). Resistance of cancer cells to immune recognition and killing. *Medical Hypothesis*, **54**(3), pp.456-460.

47   National Cancer Institute. (2010). Chemicals in Meat Cooked at High Temperatures. Retreived from http://www.cancer.gov/cancertopics/factsheet/Risk/cooked-meats

48   Cross, A.J., & Sinha, R. (2004). Meat-related mutagens/carcinogens in the aetiology of colorectal cancer. *Environmental and Molecular Mutagenesis*, **44**(1), pp.44-55. Cross, AJ *et al*. (2005). A prospective study on meat and meat mutagens and prostate cancer risk. *Cancer Research*, **65**(24), pp.11779-11784. Knize MG, Felton JS. (2005). Formation and human risk of carcinogenic heterocyclic amines formed from natural precursors in meat. *Nutrition Reviews*, **63**(5), pp.158–165.

49   Frandsen, H *et al*. (2008). Bio-monitoring of Urinary metabolites of 2-amino-1-methyl-6-phenylimidazo [4, 5-b]pyridine (PhIP) following human consumption of cooked chicken. *The Journal Food and Chemical Toxicology*, **46**(9), pp.3200-3205.

50   Steck, S.E *et al*. (2007). Cooked Meat and Risk of Breast Cancer: Lifetime versus Recent Dietary Intake. *Epidemiology*, **18**, pp.273-382.

51   Cross, "Meat-related mutagens/carcinogens in the aetiology of colorectal cancer", 2004.

52 Ibid.

53 Rohrmann, S *et al*. (March, 2009). Dietary intake of meat and meat-derived heterocyclic aromatic amines and their correlation with DNA adducts in female breast tissue. *The Journal of Mutagenesis*, **24**(2), pp.127-132.

54 Chen, J.W *et al*. (2012). Carcinogenic potencies of polycyclic aromatic hydrocarbons from back-door neighbours of restaurants with cooking emissions. *Science of the total environment*, 417-418, pp.68-75. doi: 10.1016/j.scitotenv.2011.12.012.

55 Lauber, S.N., *et al*. (2004). The cooked food derived carcinogen 2-amino-1-methyl-6-phenylimidazo[4,5-b] pyridine is a potent oestrogen: a mechanistic basis for its tissue-specific carcinogenicity. *Carcinogenesis*, **25**(12), pp.2509-2517.

56 Ibid.

57 DeBruin, L.S *et al*. (2001). Detection of PhIP (2-amino-1-methyl-6-phenylimidazo [4, 5-b] pyridine) in the milk of healthy women. *Chemistry Research Toxicology*, **14**, pp.1523-1528.

58 Rohrmann, "Intake of meat and meat-derived heterocyclic aromatic amines", 2009.

59 Lauber, S.N., *et al*., ibid.

60 Greger, M. (January 18th, 2013). *PhIp: The Three Strikes of Breast Carcinogen*. Volume 12. Retreived from http://nutritionfacts.org/video/phip-the-three-strikes-breast-carcinogen/

61 Ibid.

62 Greger, *"The Three Strikes of Breast Carcinogen"*, 2013.

63 Ibid.

64 Punnen, S *et al*. (2011). Impact of Meat Consumption, Preparation, and Mutagens on Aggressive Prostate Cancer. PLoS One, **6**(11): e27711.

65 Kabat GC, Cross AJ, Park Y *et al*. (2009). Meat intake and meat preparation in relation to risk of postmenopausal breast cancer in the NIH-AARP diet and health study. *International Journal of Cancer*, **124**(10), pp.2430–2435. Rodriguez C, McCullough ML, Mondul AM *et al*. (2006). Meat consumption among Black and White men and risk of prostate cancer in the Cancer Prevention Study II Nutrition Cohort. *Cancer Epidemiology, Biomarkers and Prevention*, **15**(2), pp.211–216.

66 DeBruin, "Detection of PhIP (2-amino-1-methyl-6-phenylimidazo [4, 5-b] pyridine) in the milk of healthy women", 2001. Greger, *"The Three Strikes of Breast Carcinogen"*, 2013.

67 Zheng W *et al*. (Nov 18th, 1998). Well-done meat intake and the risk of breast cancer. *Journal of the National Cancer Institute*, **90**(22), pp.1724-1729.

68 Sellman, S. (1999). The Hormone Hersey Supplement, www.sellman.com

69 Women's Health Care Physicians. (June, 2014) Committee Opinion. Tamoxifen and Uterine Cancer. *The American Collage of Obstetricians and Gynecologists*, Number *601*. Sismondi, P., Biglia, N., Volpi, E., Giai, M., de Grandis, T. (1994). Tamoxifen and endometrial cancer. *Ann NY Acad Sci*, *734*, pp.310–21. Fisher, B., Costantino, J.P., Redmond, C.K., Fisher, E.R., Wickerham, D.L., Cronin, W.M. (1994). Endometrial cancer in tamoxifen-treated breast cancer patients: findings from the National Surgical Adjuvant Breast and Bowel Project (NSABP) B-14. *J Natl Cancer Inst*, *86*, pp.527–37. Davies, C., Pan, H., Godwin, J., Gray, R., Arriagada, R., Raina, V et al. (2013). Long-term effects of continuing adjuvant tamoxifen to 10 years versus stopping at 5 years after diagnosis of oestrogen receptor-positive breast cancer: ATLAS, a randomised trial. *Adjuvant Tamoxifen: Longer Against Shorter (ATLAS) Collaborative Group [published erratum appears in Lancet*, [381:804]. Lancet, pp.381, pp.805-816.

70 Ibid.

71 Cohen, R. (November 1st, 1997). *Milk: The Deadly Poison* (1st ed). Published by Argus Publishing.

72 Epstein, S.S. (1996). Unlabeled milk from cows treated with biosynthetic growth hormones: a case of regulatory abdication. *International Journal of Health Services*, *26*(1), pp.173-185.

73 Epstein, S.S. (1990). Potential public health hazards of biosynthetic milk hormones. *International Journal of Health Services*, *20*(1), pp.73-84. Epstein, S.S. (1990). Questions and answers on synthetic bovine growth hormones. *International Journal of Health Services*, *20*(4), pp.573-82.

74 De Leon. D.D., Wilson, D.M., Powers, M., & Rosenfield, R.G. (1992). Effects of insulin-like growth factors (IGFs) and IGF receptor antibodies on the proliferation of human breast cancer cells. *Growth factors*, *6*, pp.327-336.

75 Peyrat, J.P et al. (1993). Plasma insulin-like growth factor 1 (IGF-1) concentrations in human breast cancer. *Eur J Cancer*, *29A*(4), pp.492-497.

76 Musgrove, E.A et al. (1993). Acute effects of growth factors on T471 breast cancer cell cycle progression. *Eur J Cancer*, 29A(16), pp.2273-2279. Heldrin, C.N., and Westermark, B. (1984). Growth factors mechanism of action and relation to oncogenes. *Cell*, *37*(1), pp.9-20. DOI: 10.1016/0092-8674(84)90296-4.

77 Macaulay, V.M. (1992). Insulin-like growth factors and cancer. *Br J Cancer*, *65*, pp.311-320.

78 Outwater, J.L et al. (1997). Dairy products and breast cancer; the IGF-1 estrogen, and bGH hypothesis. *Medical Hypothesis*, *48*, pp.453-561.

79 The European Commission. *Health and Consumer Protection. Scientific Committee on Veterinary Measures relating to Public Health: Outcome of discussions.* http://europa.eu.int/comm/dg24/health/sc/scv/out19_en.html

80 Epstein, "Potential public health hazards of biosynthetic milk hormones", 2009. Xian, C.J et al. (1995). Degradation of IGF-1 in the Adult Rat GIT in limited by a specific antiserum of the dietary protein Casein. *Journal of Endrocrimology*, **146**, pp.215-225. doi: 10.1677/joe.0.1460215.

81 Weed, S.S et al. (1996). *Breast Cancer? Breast Health!* Ash Tree Publishing, Woodstock, New York, p.203. Rinzler, C.A et al. (1996). *Estrogen and Breast Cancer.* Hunter House, California, pp.148-149. De Gregorio, W.M, Wiebe, J.V (1999). Tamoxifen and Breast Cancer. Published by Yale University Press. Study reaffirms Tamoxifen's dark side. *Science News*, p.356. Sellman, S. (1997). Hormone Heresy: What Women MUST Know About Their Hormones. *GetWell International*, USA, pp.107-108. Science News. (February 26, 1994). Studies spark Tamoxifen controversy, p.133.

82 Ibid.

83 Anderson M. (2009). Healing cancer from inside out: No animal foods, pp.85-96. Published by www.RaveDiet.com

84 Ibid.

85 Willett, W.C et al. (2005). Diet and cancer: an evolving picture. *JAMA*, **293**, pp.233-234. Negri, E. (2000) Red meat intake and cancer risk: a study in Italy. *Int J Cancer*, **86**, pp.425-428. Kaaks, R., & Riboli, E. (1997) The role of multi-centre cohort studies in studying the relation between diet and cancer. *Cancer Lett*, **114**, pp.263-270. Thorogood, M., Mann, J., Appleby, P., McPherson, K. (1994). Risk of death from cancer and ischaemic heart disease in meat and non-meat eaters. *Br Med J*, **308**, pp.1667-1670. Chang-Claude, J., Frentzel-Beyme, R., & Eilber, U. (1992). Mortality patterns of German vegetarians after 11 years of follow-up. *Epidemiology*, **3**, pp.395-401. World Cancer Research Fund. (2007). *Food, nutrition, physical activity, and the prevention of cancer: A global perspective.* American Institute of Cancer Research. Washington, DC. Sinha, R., Rothman, N., Brown, E.D et al. (1995). High concentrations of the carcinogen 2-amino-1-methyl-6-phenylimidazo-[4,5] pyridine [PhIP] occur in chicken but are dependent on the cooking method. *Cancer Res*, **55**, pp.4516-4519; Armstrong, B., & Doll, R. (1975). Environmental factors and cancer incidence and mortality in different countries, with special reference to dietary practices. *Int J Cancer*, **15**, pp.617-631.

26

The Health Consequences of Pesticidal Residues

There is rising concern about pesticides used on plants for food causing endocrine disruption, meaning that the residual pesticides appear to be changing the hormone levels in our populations.

Daniel G. Amen
Change Your Brain, Change Your Body: Use Your Brain to Get and Keep the Body You Have Always Wanted

In 1962, the book "Silent Spring" by biologist Rachel Carson was published. It described the environmental impacts of indiscriminate dichlorodiphenyltrichloroethane (DDT) spraying in the United States. Rachel Carson clearly questioned the logic of releasing large amounts of carcinogenic chemicals into the environment without a sufficient understanding of their effects on ecology or human health.

The book established that DDT and other pesticides had been shown to cause cancer, and that their agronomic use was a threat to wildlife, particularly birds. Its publication was a seminal event as regards the environmental movement and resulted in a large public outcry – that in time – led to a ban on the agricultural use of DDT in the United States and, eventually, in other countries throughout the world.[1]

Rachel Carson was on the right track when she mentioned:

Harm may be done by two or more different carcinogens acting together, so that there is a summation of their effects. If a person is exposed to DDT, for example, is almost certain to be so exposed to other liver-damaging hydrocarbons, which are so widely used as solvents, paint removers, degreasing agents, dry-cleaning fluids, and anesthetics. Cancer is a complicated disease. The fact that one chemical may act on another to alter its effect. Cancer may sometimes require the complementary action of two chemicals, one of which sensitizes the cell or tissue so that it may later, under the action of another or promoting agent, develop true malignancy.[2]

As already mentioned in previous chapters, cancer needs to be initiated and then promoted to have any long-term harmful effects on the human body. One or many carcinogens can play their role in the initiation stage of the disease, and the promotion stage is when cancer actually shows its true colours. This is when cancer grows, develops and may then progress into the metastatic (progression) stage of the disease. DDT is one of the many few carcinogens that may play a vital role in the initiation stage of cancer and many other degenerative diseases.

However, DDT was eventually banned in 1972 and replaced by the pesticide Dieldrin – subsequently also found to be so toxic that it was outlawed in 1974. Nevertheless, Dieldrin it is still around and may appear to be the reason why every single prospective study on the link between dairy consumption and Parkinson's disease, concludes that the more milk is consumed, the greater is the number of people with Parkinson's.[3] Though Dieldrin has been banned, people continue to be exposed to the pesticide through contaminated dairy products and meats; due to the persistent accumulation of the pesticide in the environment.

What else can these persistent pollutants do? Research published over the last 5 years shows that pesticides are endocrine-disrupters, and can also increase the risk of endometriosis, fibrosis, diabetes, hypertension, cardiovascular disease mortality, gum diseases, and babies having smaller brains, lower intelligence, poor attention span and other cognitive impairments, including paediatric respiratory infections.[4]

Pesticides and other chemical toxins can also raise oestrogen levels because these chemicals mimic sex hormones once they are inside the body. Since 95 percent of the pesticides Americans consume come from meat and dairy products,[5] today's average American is double-boosting his hormone levels, when he or she, consumes animal products.

The recommended daily portions of fruits and vegetables we have been advised to consume should be around 5-9 servings per-day. Let me make one thing clear though, before we deal with the number of servings of fruits and vegetables.

First, I need you to understand a very important issue concerning pesticides sprayed on fruits and vegetables.

We all know that fruits and vegetable are sprayed with pesticides, herbicides or fungicides. So, then, is it healthy and worth consuming so many fruits and vegetables? Should there indeed be nine servings per day?

The answer is "Yes". Even if all you have to eat is the most pesticide-contaminated plant foods: peaches, apples, tomatoes, sweet bell peppers, celery, nectarines, strawberries, blueberries and other berries, cherries, spinach, kale, grapes, potatoes and lettuce; the health benefits of this list of fruits and vegetables by far out-weighs the risks, even if you cannot find organic options. So, while organic food products are absolutely a healthier choice to make, we should never avoid buying fruits and vegetables out of fear of pesticide exposure.

According to Dr. Michael Greger:

> Not less than 800 million pounds of pesticides are used annually in the United States. Xenoestrogens (such as certain pesticides) may negatively affect male fertility, but the benefits of eating fruits and vegetables dramatically outweigh any risks of eating even conventional pesticide-laden produce.[6]

Not only do plants tend to contain significantly lower levels of industrial pollutants, because they are at the bottom of the food chain, but they also contain phytonutrients that combat the effects of some of these toxins.[7]

A group of scientists recently published a paper describing what happened when they ordered some dioxin from Dow Chemical Industry.[8]

The researchers took dioxin and some DDT, and dripped the chemicals on some human white blood cells – with and without – a variety of phytonutrients, to see which of these plant-compounds could have a protective role. They identified 2 as the most effective agents in protecting human blood cells from developing these cell toxic effects of dioxin and DDT. The best phytonutrients in reducing the risk of human toxicity and inflammation where found to be zerumbone, found in ginger, and auraptene, found in citrus fruits. One must keep in mind that these tests were obviously very preliminary. They were performed in a test tube and they only tested a minute fraction of the tens of thousands of known phytonutrients that can be found in the plant kingdom.

However, the results of the study do open up the possibility that plant-based diets may play a dual role in protecting us against industrial pollutants, reducing exposure, and potentially reducing some of the damage done from any chemical we are still exposed to – considering the extent of pollution to which our world has reached.[9]

Animal Products Actually Contain the Most Pesticide Residues[10]

Animal products like beef, chicken and pork were found to contain higher levels of pesticides than any other plant food. Pesticides are deposited and stored in the fatty tissues of animals. So, what ends up in them, ends up on your dinner table! When these animals are fed with pesticide-laden feed, it carries over to the finished product.

Scientists have theorized that when plants are grown without pesticides, they are forced to deal with the stress of insects, which causes them to produce more of those wonderful phytonutrients (antioxidants) compounds that are extremely vital towards human health.[11]

Pesticide residues in fruits and vegetables are actually present in trace amounts and no study – which I know of have ever linked or established a correlation between eating mainly plant foods, these residues and cancer. The opposite is true: the consumption of vast amounts of fruits and vegetables has constantly been linked to a decrease in the risk of developing cancer. There is no doubt that the benefits associated with an increase intake of plant foods; exceed many times over any hypothetical negative effects that trace amounts of contaminants might cause.

Buying organic plant-based foods are clearly the healthiest choice. Organic foods taste better, and organic agriculture protects farmers, the environment, our animal friends, including ourselves. If you cannot buy organic products, then it surely makes sense to peel fruits whenever possible, and not eat potato skins. As already mentioned in a previous chapter, it is important to remember to remove and discard the outermost leaves of lettuce, cabbage and other vegetables (if not organically grown), as these contain the most pesticide residues. Other surfaces that cannot be peeled can be washed and cleaned with a mixture of vinegar and lemon juice or with a commercial vegetable wash.

I personally thoroughly wash and do not peel the skins of the fruits I buy, as the benefits and concentration of nutrients and antioxidants are mostly found in the peels of fruits that, I believe, out-weigh the risks. It is your choice. Peel or not peel, you choose.

Finally, with deep respect to the late Rachel Carson, here are some of her most inspiring writings and quotes:

- For the first time in the history of the world, every human being is now subjected to contact with dangerous chemicals, from the moment of conception until death
- But man is a part of nature, and his war against nature is inevitably a war against himself.
- The more clearly we can focus our attention on the wonders and realities of the universe about us, the less taste we shall have for destruction.
- Only within the moment of time represented by the present century has one species – man – acquired significant power to alter the nature of his world.
- Can anyone believe it is possible to lay down such a barrage of poisons on the face of the Earth without making it unfit for all life? They should not be called "insecticides", but "biocides".

Notes for chapter 26

1. Carson, R. (1962). *Silent Spring*. Houghton Mifflin, first edition.

2. Carson, R. (2000). *Silent Spring* (Penguin Modern Classics):One in Every Four, **14**, pp.209. Published by Penguin Classics; New Ed edition (1000).

3. Kanthasamy, A.G., et al. (2005). Dieldrin-induced neurotoxicity: relevance to Parkinson's disease pathogenesis. *Neurotoxicology*,**26**(4): pp.701-719.

4. Greger, M. (December 10th, 2009). Avoiding other Banned Pesticides. Industrial pollutants in the food supply may help explain the link between dairy consumption and Parkinson's disease. Volume 3 video. Retrevied from http://nutritionfacts.org/video/avoiding-other-banned-pesticides/

5. Robbins, J. (1998). *Diet for a New America: How your food choices affects your health, happiness and the future of life on Earth* (2nd ed). Published by HJ Kramer.

6. Greger, M. (November 2nd 2012). *Plants vs.Pesticides*. Volume 11. Retrevied from http://nutritionfacts.org/video/plants-vs-pesticides/

7. Ahmed, KS et al. (May, 2013). Pesticides-induced oxidative stress: Possible in vitro protection of antioxidants. *Journal of Toxicology and Evironmental Health Sciences*, **5**(5), pp.79-85. Saxena R, Garg P, Jain DK (2011). In vitro anti-oxidant effect of vitamin E on oxidative stress induced due to pesticides in rat erythrocytes. *Toxicol Int*, **18**, pp.73-76.

8. Sciullo, E.M et al. (2010). Effects of selected food phytochemicals in reducing the toxic actions of TCDD and p,p'-DDT in U937 macrophages. *Arch Toxicology*, **84**(12),pp.957-966.

9. Greger, "Plants vs. Pesticides", 2012.

10. Ibid.

11. Bagchi K, Puri S. (1998). Free radicals and antioxidants in health and disease. *East Mediterranean Health Jr*, **4**, pp.350–60. Sies H. (1997). Oxidative stress: Oxidants and antioxidants. *Exp Physiol*, **82**, pp.291–295. Magnenat JL, Garganoam M, Cao J. (1998). The nature of antioxidant defense mechanisms: A lesson from transgenic studies. *Environ Health Perspect*, **106**, pp.1219–28. Matill HA. (1947). Antioxidants. *Annu Rev Biochem*, **16**, pp.177–92. Temple, N. J. (2000). Antioxidants and disease: More questions than answers. *Nutr. Res*, **20**, pp. 449-459. Jie, S et al. (2002). Antioxidant and Antiproliferation Activities of Common Fruits. Department of Food Science and Institute of Comparative and Environmental Toxicology, Stocking Hall, Cornell University, Ithaca, New York 14853-7201.

27

Carnism: Loving Some Animals and Eating Others

I do not eat anything with a face and if slaughterhouses had glass walls, everyone would be a vegetarian.

The late Linda Louise, Lady McCartney (1941-1998)
An American musician, photographer, animal rights activist and was married to Sir Paul McCartney, a founding member of the Beatles

Carnism: The Invisible Belief System

Scientific literature and studies carried out on many different types of foods have proven that the impact of a plant-based diet on human health is significant. But our diet does not affect only our personal health: it raises moral issues about our relationship with animals, particularly those we have "domesticated" for our own selfish needs and those we capture in the wild for other purposes.

Carnism is the invisible belief system, or ideology, that conditions people to eat certain animals. Most people vision eating animals as a given, rather than a choice. In meat-eating cultures around the world, people typically don't think about why they find the flesh of some animals disgusting and the flesh of other animals appetizing, or why they eat any animals at all. But when eating animals is not a necessity for survival, as is the case in much of the world today, it is a choice – and choices always stem from beliefs.[1]

It is strange how many people care so much about animals, yet still decide to eat many so-called domesticated animals that are raised for food and are being slaughtered every single second, minute, day after day, without realizing the effect that this has on human health itself, the environment, and the species of animals we actually call "our friends". Do you not often wonder why we still call some animals "pets" and lavish our love and care for them, and then turn around and call other animals "dinner" and allow them to be treated as if they had no feelings or needs of their own.[2]

Why Eating Animals is a Social Justice Issue

The invisibility of carnism makes eating animals appear to be simply a matter of personal ethics, rather than what it actually is: the inevitable end result of a deeply entrenched, oppressive system. Carnism is structured like other "isms," such as racism, sexism, and heterosexism, which are organized around the oppression of certain groups of "others." And while the experience of each set of victims will always be somewhat unique, the ideologies themselves are structurally similar, as the mentality which enables such oppression is the same.

If we fail to pick out the common threads that are woven through all oppressive systems, then, we will simply trade one form of oppression for another. Thus, to create a more humane and just society, we must include carnism in our analysis.³

Most of us who have grown-up eating animals do not realize that every time we sit down to a meal, we are acting in accordance with an invisible belief system: that has shaped our thoughts, preferences, feelings and behaviors. We aren't aware of how we have been conditioned to eat animals without considering the implications of our choices on ourselves or on others – or even realize we are making choices at all. This invisible belief system, carnism, has created the illusion that when we eat animals we are making our choices freely. But carnism is structured to enable humane people to participate in inhumane practices; without realizing what they are doing; to block our awareness; so that we unknowingly act against our own interests, and the interests of others. If you eat animals, you need and deserve to know the truth about carnism, so that you can make your choices freely, because *without awareness, there is no free choice.*⁴

The Exploitation of Domesticated Farm Animals

Many of us still envisage images of farm animals tended by cheerful farmers over vast tracts of rolling pasture land, with chickens and turkeys roaming freely around the barn. We think of poultry peeking at fresh grain and laying eggs at their whim, while the family cow wanders near the hen-house, silver bell tied around its neck clanging lazily, and pigs enjoying themselves rolling in a pool of mud, eating to their heart's content the feed provided by their friendly farmer.⁵

This may have been true, once. But the real truth now is that all this has changed. This is not what exists today. In most westernized countries billions of farm animals are slaughtered every year for human needs, all hidden away from our eyes in massive factory or battery farms.⁶

The reality is that most people only consider farm animals as a meal on their plate and nothing more, even though they still say that they respect animals and have their welfare at heart. Why respect birds, dogs or cats, and not chickens, cows or pigs? It is all about tradition which has now turned into a so-called "normal eating culture".[7]

We have also been wrongly taught that animal produce is needed in a human diet: for protein, calcium and other energy or nutrient values. This certainly is not true.

We know now that we can get all our required nutrients by just eating plants. In eating animal flesh the end result is that the animals we are slaughtering for food are actually killing *us*. This might sound "funny", but animal flesh and dairy products will, in the long run, get their revenge, causing many humans chronic degenerative diseases.[8]

Today's farming practices have indeed changed. Why? For the sake of maximizing profits and streamlining production, most of today's animal products originate in factory-style farms that are massive industrial warehouses: crammed with thousands of chickens, turkeys, cows, pigs, and other animals. Suppliers strive to maximize "efficiency" by raising huge numbers of these animals in the shortest amount of time. Sadly, the rule is, mass produce for max profits![9]

In the U.S.A., in order to shorten the growth period, most factory farms employ artificial "growth promoters", such as Recombinant Bovine-Growth Hormone (rBGH). These injected hormones allow animals to develop muscles and milk much faster than normal. Although the European Union has banned the use of artificial hormones, approximately two thirds of all beef cattle in the U.S. are treated with them. In fact, the U.S. is the only developed nation that permits the use of rBGH for increasing animal flesh and milk production in dairy cows.[10]

Factory farm animals in the U.S and similarly in other western nations, are being pushed to their biological limits and subjected to tremendous amounts of stress. Milk cows have to endure gruelling demands that cause disease and fatigue. Most cows today are forced to produce nearly 100 lbs of milk every day, which is ten times what they produce naturally. The normal life-span of cows is around 20 years, but many factory-bred cows can barely walk by their fourth birthday!

> **FACT BOX**
>
> **Length of time that baby calves will suckle from their mothers in a natural situation:** *8 months.*
>
> **Age at which U.S. dairy calves are routinely taken from their mothers and transported to veal stalls:** *less than 24 hours.*
>
> **Percentage of U.S. dairy calves taken from their mothers within 24 hours of birth:** *90 percent.*
>
> Retreived from Robbins J.(2001). *The Food Revolution*, pp.188-189. Published by Conari Press.

Hens are also exploited for egg production. Hens and chickens are commonly packed in cages so tightly that they can hardly move or flap their wings. Chicks' beaks are removed (a process called "debeaking") so that they do not peck each other, as when too many of them are confined to overcrowded cages they can peck and fight each other to death. Chickens are hung by their feet with shackles; because poultry is not covered by the Humane Methods of Livestock Slaughter Act, there is no legal requirement to stun them before slaughtering.[11]

Calves raised for veal are separated from their mothers immediately after birth to stop them from suckling on their mother's udders, so that the milk is collected for human consumption. Calves are then confined thereafter to crates measuring just two feet wide, their movements restrained with neck chains; male calves are castrated without painkillers.[12]

The cost of every square inch of space is carefully calculated to maximize profits.

Pigs, too, are packed tightly together, and made to constantly breathe the obnoxious gases released from their own excrement.[13]

It is sad, but this is the reality of how some nations treat factory-farm animals, and one of the few reasons why I personally decline to eat animal food. Even if animals were treated humanely, I would still not kill any animal for food. I consider it to be immoral and unnecessary to inflict pain on any animal for no reason at all, especially when they cannot defend themselves. We can live healthier lives without eating animal flesh and without consuming dairy products.

Charles Darwin (1809-1882) spoke out and said: "Like humans, the animals feel joy and pain, happiness and unhappiness".

These are the fundamental and ethical reasons why many people turn to a vegetarian or vegan diet. The cruelty inflicted on animals is indeed immoral. As intelligent human beings should we not be showing more compassion? Instead, we slaughter the weakest and most defenseless animals for our own self-seeking contentment.

The Indisputable Reasons *why we all should stop Consuming Animal Products*

But if animal cruelty is not enough of a valid reason for you to stop eating animal products, then there are other earth-worthy responsible reasons:

1. Collectively, farm animals themselves are responsible for a fifth of all human-induced greenhouse gases (Methane and carbon emissions that result from the massive infrastructure required to transport livestock).
2. Massive land areas equivalent to 7 football fields are destroyed in the Amazon basin *every minute* for harvesting soybeans, wheat and corn to feed worldwide livestock.
3. In a 1997 report by the U.S. Senate Agriculture Committee, it was said that animals raised for slaughter produce 130 times as much waste as the entire human population. Human waste is chemically treated in sanitation plants, but animal waste is not. Animal waste is sprayed onto land, and then much of it runs off to pollute groundwater, rivers or streams.
4. The U.S. Environmental Protection Agency estimates that 1lb of processed beef requires 2,500 gallons of water.
5. Much of the land harvested exclusively for animal feed is saturated with pesticides and fertilizers used to grow crops as rapidly as possible. These chemicals seep into groundwater and spill into rivers and oceans, depleting oxygen in the water, killing fish and other marine life.
6. On top of all that, livestock waste further contaminates the water supply with hormones and antibiotics excreted by the animals.

These are the facts and they are indisputable. What more can we do? The number of farm animals slaughtered for food worldwide is growing, and the treatment of these animals has become increasingly merciless.[14]

The healthy option is to increase harvest and cultivating plant foods, instead of animal products. Plant foods, such as green-leafy vegetables, fruits, legumes, nuts, seeds and grains are more efficient to produce than animal foods. The net result would be that more food will be available to developing or undeveloped countries to feed the poor, sick, or hungry sections of their populations.

The human world population is forecast to rise from 6.8 billion to 9.1 billion by the year 2050. The more incomes rise, the more people eat animal products. The Food and Agriculture Organization (FAO) has forecast that by 2050 meat consumption will more than double, while milk consumption will rise by around 80 percent.[15]

The destruction of mankind will be inevitable if we do not change our eating habits. This is why many people around the world are vegetarian or vegan. The transition to a plant-rich diet will unquestionably reduce the suffering of farm animals, reduce global warming and deforestation, resolve global issues concerning waste and water pollution, soil erosion, and eliminate the extinction of endangered species.[16] Lastly, and more significantly, it may also reduce mankind's health risk of developing non-communicable diseases.

Ralph Waldo Emerson (1802-1882), many years ago used to say: "You have just dined and, no matter how far the distance, you are complicit in the original murder".

I was really emotional stuck when I first read Emerson's statement. Oh my God, he was right! It is so easy for us today to forget we are consuming meat. Most people will adopt a vegetarian lifestyle if they knew they had to kill animals to feed on it. Deep down people will not admit that they are killing an intelligent animal that was alive before it is slaughtered, processed and brought to their local butcher's or supermarket.

From the moral point of view I personally refuse to eat meat for this selfsame cause. We do not have to slay any type of animal to sustain human health as we can get all our adequate amounts of essential nutrients by just eating a whole spectrum of plant foods, with the exception of just one vitamin – Vitamin B_{12} (read chapter 8 for alternatives).

Martin Luther King Jr was quoted as saying:

"One day the absurdity of the almost universal human belief in the slavery of other animals will be palpable. We shall then have discovered our souls and become worthier of sharing this planet with them"

And according to Abraham Lincoln, who merely said:

"I am in favor of animal rights as well as human rights. That is the way of a whole human being."

For our health, for animal welfare, and for the impact that animal products leave on the Earth's environment, I will *never* eat the flesh of any creature that lives or used to live on planet Earth!

Please, before you eat animal foods, do consider your responsibilities – especially towards animal rights, world population health and for the well-being of the ecosystem. Disregarding all this is not only selfish, but also ethically immoral. Many of these issues can be resolved if more people transition to a plant-based diet. It is *that* simple!

Crucially, and not forgetting – we must all try to fully understand the important aspects of Carnism and try to abolish this invisible belief system, or ideology, that conditions people to eat certain animals and cherish others. For all moral human beings there may perhaps be someday a light of hope at the end of a long dark-weary-tunnel. This new tunnel may someday ultimately lead, to a road full of hope, which rightfully challenges Carnism and the philosophy behind it.

Notes for chapter 27

1. What is Carnism? CAAN Carnism awareness and action network. Retreived from http://carnism.org/

2. Melanine Joy. (Feb 27th, 2012). Carnism: The Pyschology of Eating Meat. Retreived from https://www.youtube.com/watch?v=7vWbV9FPo_Q

3. What is Carnism? CAAN Carnism awareness and action network: Why eating animals is a social jusice issue? Retreived from http://carnism.org/2012-05-09-15-00-33

4. Ibid.

5. Robbins, J. (2001). The Food Revolution: How your diet can help save your life and our world. Chapter 10: Old McDonald Has a Factory, pp.165-183. Published by Conari Press. Scully, M. (October 1st, 2003). *Dominion: The Power of Man, the Suffering of Animals, and the Call to Mercy* (1st ed). Published by St. Martin's Griffin.

6. Eisnitz, A. Gail. (Nov 1st, 2006). *Slaughterhouse: The Shocking Story of Greed, Neglect, and Inhumane Treatment inside the U.S. Meat Industry*. Published by Prometheus Books.

7. Joy, M. (2011). *Why We Love Dogs, Eat Pigs, and Wear Cows: An Introduction to Carnism*. Published by Conari Press.

8. Ibid.

9. Eisnitz, *"Slaughterhouse"*, 2006.

10. Joy, *"Why We Love Dogs, Eat Pigs, and Wear Cows"*, 2011.

11. Becker, S.G. (March 24th, 2009). Congressional Research Service. Nonambulatory Livestock and the Humane Methods of Slaughter Act: Nonambulatory Diabled Cattle: pp.1-7. Retreived from http://nationalaglawcenter.org/wp-content/uploads/assets/crs/RS22819.pdf

12. Robbins, J. (2012). *No Happy Cows: Dispatches from the Frontlines of the Food Revolution*. Published by Conari Press.

13. Ibid.

14. Robbins, *"No Happy Cows"*, 2012.

15. The Food and Agriculture Organization. How to Feed the World in 2050. Retreived from http://www.fao.org/fileadmin/templates/wsfs/docs/expert_paper/How_to_Feed_the_World_in_2050.pdf

16. Hoekstra, A.Y *et al.* (2012). The hidden water resource use behind meat and dairy. *Animal Frontiers*, **2**(2), pp.3-8. Shepherd, J.M. (July, 2011). Carbon, climate change, and controversy. *Animal Frontiers*, **1**(1), pp.5-13. Thomas F. Ducey, Anthony D. Shriner, Patrick G. Hunt. (2011). Nitrification and Denitrification Gene Abundances in Swine Wastewater

Anaerobic Lagoons. *Journal of Environment Quality*, **40**(2), pp. 610 DOI: 10.2134/jeq2010.0387.? Marlow, H.J *et al.* (May, 2009). Diet and the environment: does what you eat matter ? *Am J Clin Nutr*, **89**(5), pp.1699S-1703S. Baroni L, Cenci L, Tettamanti M, Berati M. (February, 2007). Evaluating the environmental impact of various dietary patterns combined with different food production systems. *Eur J Clin Nutr*, **61**(2), pp.279-86.

28

Rebounding: Cleansing and Boosting the Human Immune System Against Cancer

Lack of activity destroys the good condition of every human being, while movement and methodical physical exercise save it and preserve it

Plato (428 B.C.-348 B.C.)
Ancient Greek philosopher

A Rebounder is an Exercise Equipment that can be Effective Against Cancer

A rebounder is an exercise equipment that works like a trampoline: you slowly, safely, and steadily, bounce up and down.

Simply put, bouncing on a rebounder approximately two minutes for up to 1 hour per day is an ideal management protocol for defending against cancer. Within just two minutes the entire lymphatic system is enhanced, creating a demand for more lymphocytes, including killer T-cells. The white blood-cell count is approximately tripled during this two-minute rebounding session. Many cancer-killing potions (such as hydrogen peroxide) are produced more efficiently, whenever a patient is rebounding.

Approximately one hour after rebounding, the white blood-cell count returns to normal, so it is advisable to rebound again to flush out the lymphatic system and to create another army of defensive immune cells. The object of this pattern is to keep the immune system operating optimally and healthy logically, hour after hour, day after day. Do this until it is time to eat, rest or sleep.

Rebounding exercise also strengthens each cell of the body, so that healing can occur. Every process of every cell is enhanced, allowing each organ to do its job more efficiently: digestion and absorption are enhanced; elimination is improved, thus removing more toxins from the body. The body is relaxed and de-stressed, and healing chemicals are produced. A stabilizer bar should be available and attached to the rebounder: to add security, safety, balance, and support whenever needed, especially for the elderly or for patients who are considered weak or unsteady. If endurance is low, start with less time than two minutes and try to work up to the optimum two minutes over a period of a few weeks.[1]

Because rebounding is a compact force on the cells of the body, finding your threshold for health bouncing is important in order to avoid over-exerting, and safeguarding yourself. This is done by timing your initial two-minute bouncing session with the second hand of a watch or clock. If dizziness, pain, weakness, or other discomfort occurs before the two minutes have elapsed, that is the time to stop bouncing, noting the length of time that was bounced. Whatever the time frame, from a few seconds to two minutes, that is your threshold for health bouncing; that advisable time frame should be rebounding for at least 1 hour each day, and to gradually work up to the two-minute session or more, according on how healthy and comfortable you feel. Never over exert yourself. If the person is too weak to stand, a "buddy bounce" can be employed. Being creative in finding a way to health bounce is important for those who are weak or have poor balance.[2]

Blood-Oxygen Levels are Enhanced via Rebounding

It is important to remember to breathe deeply while bouncing, and to exhale as completely as possible, making room for more air with your next breath. As you rebound and breathe deeply, you can also train yourself to breathe properly; when you are not on the rebounder.

Oxygen in the bloodstream plays an important role in cancer suppression. Previously mentioned in this book, Dr. Otto Warburg, a Nobel Prize winner, demonstrated during the 1920s that the metabolism of cancerous tissue differs radically from that of normal tissue. Normal cells and tissues need adequate amounts of oxygen to carry out aerobic respiration; however, cancerous tissue needs little or no oxygen to strive. This is substantiated by the fact that malignant tumours are frequently found in areas in the body where the oxygen blood supply is low.[3]

An abundant supply of oxygen to the cells seems to be a definite defense against cancer. Also, because rebounding exercises are specific types of oxygenating exercises, regular bouncing throughout the day, supplies an increased amount of oxygen to all the cells of the body simultaneously. At the same time, rebounding also helps to boost our immune system and removes harmful-toxins from the body.

Consciously making an effort to breathe deeply while you bounce; delivers an even greater supply of oxygen to the bloodstream. Rebounding, efficiently enhances oxygen levels and nutrients to every cell – removing toxins – waste products – dead cancer cells, which will, overall enhance the body's healing process.

According to Linda Brooks, who healthy advises:

It is important and necessary to educate oneself about standards for purchasing a rebounder; because all rebounders are not created equal. In her experience many rebounders on the market today can actually be harmful to one's health, so be sure to choose carefully. Linda Brooks' primary choice of rebounders has always been the Needak soft-bounce rebounder, which has stood the test of time: for safety, efficiency, convenience, and cost effectiveness.

A true rebounder is not a toy. It is a piece of exercise equipment that is designed to promote the body's natural healing. Rebounding works! With every healthy bounce, each cell of the body is cleansed, strengthened, protected, and stimulated to heal itself.[4]

To make a point clear to all, that this exercise is not actually intended to help increase muscle bulk, but it is an intervention management exercise plan – especially designed and dedicated to help patients who have been diagnosed with cancer, and other chronic-degenerative diseases.

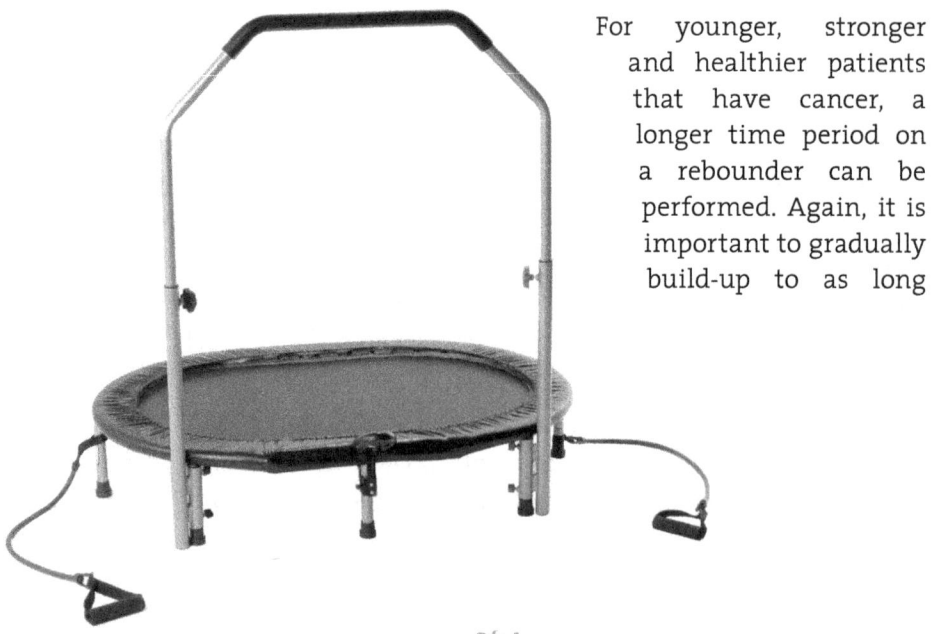

For younger, stronger and healthier patients that have cancer, a longer time period on a rebounder can be performed. Again, it is important to gradually build-up to as long as the patient feels strong and well enough. Around 10-20 minutes of rebounding per day is adequate and a good detoxification mechanism, that boosts the immune system fighting capabilities and aids in removing excessive toxins from the patient's body.

The Reduction of Muscle Loss and Osteoporosis

In 1980, the National Aeronautics and Space Administration (NASA) conducted a study in search of the best exercise to rehabilitate astronauts after spending time in gravity-deprived outer space, for periods of time.

The reason NASA carried out the study of trampolining was mostly because astronauts lost as much as 14% of their muscle mass in fifteen days of space flight. When they studied mice, with a faster metabolism, they lost as much as 40% in the same period. They were suffering from deconditioning and osteoporosis, the same noticed in older patients or the bedridden. NASA was looking for a non-traumatic form of exercise; that would help to recondition a deconditioned astronaut. Can we assume the accepted forms of exercise were not acceptable to NASA? This is the only reason they would reach out from the accepted into new and unique forms of exercise never before considered.[5] Trampoline exercises also showed better improvements on selective psychomotor variables in patients, compared to control groups.[6]

The results indicated that for similar levels of Heart Rate (HR) and Oxygen Uptake (VO2), the magnitude of the biomechanical stimuli is greater with jumping on a trampoline than with running – a finding that might help identify acceleration parameters needed for the design of remedial procedures to avert deconditioning in persons exposed to weightlessness.[7]

Physical Activity Burns Calories

All physical activity burns calories. Every movement of the body uses up some energy, even when you are less active, during sleep, as the brain is still working overtime to keep you alive. The more you move about, the more calories you burn. Also, whenever you are physically active, your body readjusts your metabolism to burn calories more quickly. This effect persists for a time, even beyond the period of activity.

You must never forget that by staying active, your muscles keep their strength. If you do not keep active, your muscles will waste away (disuse atrophy) from inactivity. This is important to note; because your muscles are real "calorie burners". Muscle tissue has a rapid metabolism and is much better than fat tissue at burning off the calories we ingest.

Physical activity also helps keep appetite under control. After a long run, or an hour on the dance floor, people are actually less likely to overeat, even though they have just burned off a lot of calories. The mechanism helps you stay slim after you have shed unwanted pounds, but unfortunately will not help as you are losing weight. While you are overweight, your appetite is not likely to be affected by your physical activity.[8]

Finally, people who use their bodies are better able to relax, and more likely to get a good night's sleep. This too is important, because when you are chronically stressed or tired, you are much more likely to use unhealthy foods, or other indulgences to prop yourself up.

Just as there are foods that naturally work to keep you slim, there are physical activities that do the same. The most important ingredient is any physical activity done as an enjoyment: if it is not fun, you will not stick to it. Physical activities do not need to be particularly strenuous, and they certainly do not have to be unpleasant. You can burn up all the gas in your tank by taking part in a high-speed chase, swim 20 laps around a swimming pool, or go for a 30-45 minute leisurely walk or stroll. Combining physical activity with a good dietary plan is all you need to lose weight naturally and stay healthy.[9]

Rebounding is also an essential part of an anticancer management program, it can also be fun. Try it!

Notes for chapter 28

1. Brooks, L. (2008). *"Rebounding for better health."* Published by Vitally Yours Press; First edition.
2. Ibid.
3. Warburg, O.H. (June 30th, 1966). The Prime Cause of Cancer & Prevention of Cancer: Respiration of Oxygen in Normal Body Cells vs. Fermentation of Sugars in Cancer Cells. The way to prevent Cancer: Revised lecture at the meeting of the Nobel Laureates at Lindau, Lake Constance, Germany.
4. Brooks, *"Rebounding for better health"*, 2008.
5. Bhattacharya, A *et al*. (Nov 1st, 1980). Body acceleration distribution and O_2 uptake in humans during running and jumping. *Journal of Applied Physiology*, **49**(5), pp.881-887.
6. Dr. S. Udhaya Shankar. (2014). Effects on selected Trampoline Exercises on Selected Psycho Variables among College Students. *International Journal of Research*, **1**(8), pp.1432-1436.
7. Bhattacharya, "Body acceleration distribution and O_2 uptake", 1980.
8. Kissileff, H.R., *et al*. (1990). Acute effects of exercise on food intake in obese and non-obese women. *American Journal of Clinical Nutrition*, **52**, pp.240-245.
9. Barnard, N. (1993). *Food for Life: How the New Four Food Groups can save your Life. Chapter 4: Real Weight Control*, pp.113-114. Published by Harmony Books.

29

Best Anti-Cancer and Anti-Angiogenic Fruits and Vegetables

They say that vegetable food is not sufficiently nutritious. But chemistry proves contrary. So does physiology. So does experience... And again: the largest and strongest animals in the world are those which eat no flesh-food of any kind – the elephant and rhinoceros.

Russell Trall (1812-1877)
An American health reformer and physician

In this chapter, we will take a closer look at some clinical studies on the anti-cancer fighting properties of some common foods of the plant kingdom: to see if "plant-based foods" can actually assist in the course of anti-proliferation, growth and in reducing angiogenesis in common known cancers.

The thousands of non-nutrient components within the plant kingdom (phytonutrients) are critical for human survival. This is because they have been found to contain miraculous, medicinal, and natural healing powers. The numerous phytonutrients and antioxidants present in a variety of plant foods is where the future of health in humans will be talked about in many generations to come. These life-saving plant chemicals are surly the answer to the future of disease prevention, and human longevity. Let us look at some of these studies to see which types of fruits and vegetables *may* actually have the capacity in fighting back against cancer.

In a study, the results of which appeared in the Journal of Food Chemistry in 2009: twenty different common vegetables were compared for their anti-proliferative and anti-cancer properties, to see which vegetables were useful against a variety of specific-cancer types.[1]

The results were amazing. They found that certain types of vegetables could actually inhibit the proliferation, growth and overall mass of certain cancer types.

The technique they used was to extract the juices from these 20 common vegetables, and place them inside Petri dishes where the different types of cancer tissues were positioned. A reaction occurred and the results were determined. They found that *most cruciferous vegetables and the alliums' family of vegetables were the best overall plant foods in anticancer inhibition*. Here are the final results according to this study. The top vegetables in each of the cancer-cell lines tested are listed below.[2]

BEST ANTI-CANCER AND ANTI-ANGIOGENIC FRUITS AND VEGETABLES

	1st	2nd	3rd	4th	5th & 6th	7th or 8th
Breast cancer	Garlic	Leeks	Onions	Brussels sprouts	Cauliflower and Cabbages	Broccoli and Kale
Brain tumours	Garlic	Leeks	Brussels sprouts	Beetroot	Cabbages and onions	Broccoli, Cauliflower and spinach
Kidney cancer	Leeks	Garlic	Cabbages	Brussels sprouts	Kale, and Spinach	Onions and broccoli
Lung cancer	Garlic	Leeks	Onions	Brussels sprouts	Kale and cabbage	Broccoli and Spinach
Medulloblastoma (child tumours)	Garlic	Leeks	Onions	Brussels sprouts	Broccoli and spinach	Cabbage
Pancreatic carcinoma	Garlic	Leeks	Cabbage	Onions	Kale and broccoli	Spinach
Prostate cancer	Garlic	Brussels sprouts	Green onions	Leeks	Broccoli and cauliflower	Kale and cabbages
Stomach cancer	Garlic	Leeks	Green onions	Yellow onions	Brussels sprouts	Cauliflower and cabbages

Anticancer Inhibition: Top 3 Vegetables in 8 Cancer-Cell Lines

	1st	2nd	3rd
Stomach	Garlic	Leeks	Green onion
Pancreatic	Garlic	Leeks	Brussels sprouts
Breast	Garlic	Leeks	Green onion
Prostate	Garlic	Brussels sprouts	Green onion
Lung	Garlic	Leeks	Green onion
Kidney	Leeks	Garlic	Cabbage
Brain	Garlic	Leeks	Green onion
Glioblastoma	Garlic	Leeks	Brussels sprouts

Based on these results, I would certainly recommend a garlic salad dressing! But, is garlic toxic to *all* cells, or is it toxic just to *cancer* cells? They even tested for this, and found that garlic actually stops the growth of tumour cancer cells but, more importantly, it actually leaves healthy human body cells intact. The same result was obtained in respect of all the other vegetables in the study. The vegetables selectively attack, and are in fact, specific in targeting cancer cells, but surprising leave normal human body cells alone. Vegetables are amazing!

If you did not pick garlic, but instead chose one of the other vegetables in the top list of anti-cancerous vegetables, you will still be reducing your risk of tumour development, and helping towards cancer prevention.

So we need to eat a portfolio of green-leafy and other types of vegetables, because they each tend to target specific or different cancers. If you are particularly concerned about a specific cancer, e.g. if you have a strong family history of breast cancer, then you may need to choose the best vegetables per-day that excel in targeting breast tissue.

The best two families of vegetables for cancer prevention in the study were:

- The cruciferous family of vegetables: broccoli, cauliflower, kale, spinach, Brussels sprouts, beetroots, red cabbage/cabbage; and
- Allium family of vegetables: garlic, onions, shallots, and leeks.

The least effective anti-cancer plant foods were:

- Lettuce and potatoes.
- Tomatoes and (believe it or not)
- Carrots.

The researchers in this study concluded that: "The inclusion of cruciferous and allium vegetables in the diet are essential for effective diet-based chemo-preventive strategies".[3]

Another conclusion was the following: "The majority of the vegetable extracts tested in this study, including vegetables that were commonly consumed in Western countries such as potatoes, carrots, lettuce and tomato, had little effect on the proliferation of the tumour cell lines."

That is why we need to select the right types of fruits and vegetables for a patient's cancer treatment.

Clinical scientific research has concluded that these anti-cancer vegetables are crucially important and without any doubt whatsoever should be included in diet management plans for breast cancer patients. The vegetables that should be included are:

- Garlic, onions, leeks and shallots
- Broccoli, Brussels sprouts, including spinach and watercress
- Red cabbage and other types of cabbages

- Cauliflower, and
- Beetroots.

I would also include and add daily additional amounts of kale, mushrooms, green tea, flax seeds, lemons and limes, orange and grapefruit pith, apples and apple peels, apricot seeds, hemp, vitamin C, vitamin D_3, and selenium for an anticancer diet.

To fight against any type of cancer, it is essential that one eats the "correct portfolio" of anti-cancer plant foods to cover our basis.

The Cancer Stem Cell: Broccoli vs. Breast Cancer Stem Cells

Over the last decade, a new theory of cancer biology has emerged "The Cancer Stem Cell."[4]

Normal Stem cells are involved in organ repair. They actually travel around the body, sit and wait, until there is some physiological damage and slowly replace whatever structures are necessary: new skin-tissue, bone, muscle, would healing, etc. Human Stem cells are always ready in the healing process of producing new cell types.[5]

However, those same qualities – migration, colonization, proliferation, self-renewal and immortality, can be used against the human body, whenever human stem cells normal functions are interrupted by a carcinogen stimuli or other toxic substances prolongs inside our bodies.

Cancer stem cells may also explain why cancer might re-emerge and metastasize, especially in those patients who are in cancer remission. There may be no cure, but only remission. You may have a breast cancer relapse 20-25 years after you thought it was surgically removed and treated with radiation, chemotherapeutic drugs, due to "Cancer Stem Cells."[6]

Our current armamentarium of chemo-drugs and radiation is based on animal models; basically, is the tumour shrinks it's considered a success. But, laboratory rats only live up to 2-3 years. So, all these new improved therapies: angiogenesis (cutting off the tumor's blood supply), are great, but the cancer stems cells can just take an alternative route.

They metastases and travel to other parts of the body and grow new tumours, which are usually more aggressive and harder to remove.

Sadly, we have known for decades that once a cancer spreads, it extremely difficult or virtually impossible to eliminate, treat or cure.

What we really need to do once you have cancer, is to strike at the root of cancer; treatments aimed not at reducing the tumour bulk, but rather on targeting the "beating heart" of the tumor, which is of course the Cancer Stem cell.

Enter Broccoli: Sulforaphane and Indole-3-Carbinol (I3C)

Breast tissue naturally has lots of stem cells; the body never knows when a woman will get pregnant and when it needs to start producing a lot of mammary glands to produce milk.

Researchers recently discovered compounds in Broccoli/Brussels sprouts; that may destroy cancerous stem cells, and prevent them in the first place from developing abnormally.[7]

Sulforaphane and I3C are found in cruciferous vegetables that are capable of detoxifying certain carcinogenic substances. Many phytochemicals found in cruciferous vegetables prevent pre-cancerous cells from developing into malignant tumours. These essential so-called anticancer chemicals also promote the suicide of cancer cells and block angiogenesis, and may increase the action of Natural-Killer Cells of the immune system against cancer (by more than 50%). Broccoli actually reduces oestrogen positive and oestrogen negative dependant tumour growth.[8]

Let's add some broccoli juice. Once the broccoli juice was added to breast cancer tumour cells in a test tube, they slowly decreased in size and completely reduced the tumour bulk. But this is in a test tube. But, how do we really know we absorb sulforaphane into our bloodstream; whenever we eat broccoli – and even if we do – how much do we have to eat; to arrive to these test tube concentrations where is counts, in breast tissue itself, where a tumour may be evolving?

An innovated group at John Hopkins University figured out this problem.⁹ They found women scheduled for breast reduction surgery, and an hour before they went into the operating room; the woman consented to drink some broccoli juice.

"In a subsequent pilot study, "Eight-Healthy Woman undergoing reduction mammoplasty" were given a dose of a broccoli sprout preparation containing 200umol of sulforaphane"[10]

On the Human Sulforaphane pharmacokinetics table, they found; an average of 2 ± 1.95 pmol/mg tissue in their left breasts and 1.45 ± 1.12 in their right.

Excitingly, now, for the first time ever, we not only know that broccoli can actually specifically target human breast cells, but we also know the final tissue concentration. We already know what broccoli does to oestrogen receptor positive and oestrogen receptor negative. So, what does it correspond to here?

To continually bath the tissues of one's breast to the concentrations used in this study, all you need to do is eat around 1¼ cups of broccoli to fully reduce breast tumour growth. In other words "It's doable". I just put them on my salad; real world effects, at real world doses. So, essentially, *"Do eat your Broccoli/Broccoli Sprouts"* in-order to help reduce breast tumour growth.

Dr. Michael Greger explains why there is *"No Sulforaphane"* found in Broccoli: only until you actually bite or chew it".

Cruciferous vegetables include: Broccoli; Brussels Sprouts; White or Red Cabbage; Cauliflower; Spinach; Kale; Turnips; Watercress; Bok Choy; Kohlrabi; Horseradish and Rutabaga, etc.. The chemicals found in cruciferous vegetables are known as "Glucosinolates" These work on releasing two classes of compounds that posse's extremely high anticancer activity: *Isothiocyanates and Indoles.*

Over one-hundred different types of glucosinolates exist in nature. Glucosinolates act as a kind of reservoir that stocks many different isothiocyanates and indoles. I will give you an example on how they work on boosting high anticancer potential.

Example: A person bites into a broccoli flower, a good source of glucosinolates. As this person crews on the broccoli, the plant cells break down and separate compartments that were present in the cells mix together.

As a consequence of the mixing the glucosinolates contained in one cell compartment, exposure to *myrosinase* is achieved: an enzyme that was present in another compartment. Myrosinase acts by breaking up parts of the glucosinolate molecules. In our particular case, chewing on the broccoli flower causes the principal isothiocyanate in this molecule, *glucoraphanin*, to find itself in the presence of myrosinase. The glucoraphanin is immediately converted into *sulforaphane*: the potential and powerful anticancer molecule.

Note: Try to avoid over boiling cabbage and broccoli. Over-boiling cooking methods can destroy or denature sulforaphane and I3C's, making these antioxidant's and other enzymatic compounds inactive.

In other words, the anticancer compounds in cruciferous vegetables: Sulforaphane, Indole-3-Carbinol (I3C) and Phenethyl Isothiocyanate (PEITC), are present in a latent state in the vegetables themselves; it is only when these vegetables are actually eaten, that the active anticancer compounds are released. These compounds can then fulfil the cancer-fighting functions they are known to accomplish.

To get the best sources of the anti-cancerous compounds in cruciferous vegetables, it is advised that you do not over-cook them, as it may denature the glucosonolates and the essential enzyme myrosinase. Both compounds should therefore be either naturally juiced or cooked as little as possible, with a minimum of liquid, to reduce the loss of myrosinase activity and glucosinolates caused by soaking the vegetable in water. "Bit-or-chew-them-well and do-not-over-cook-them"

Prevention is better than the cure. We need to stop cancer from developing in the first place and that's the *"key"* in unlocking the riddle to cancer success. Eat your cruciferous vegetables for cancer protection.

The Power of Mushrooms: Aromatase Inhibition in Breast Cancer

Breast Cancer cell's lines respond and growth well with increased oestrogen and IGF stimulation, whenever animal products are ingested, which may contain an increase of oestrogen and IGFs: Meat and dairy products may increase tissue tumour growth and oestrogen receptor positive breast cancer cells.

Most breast tumours are oestrogen receptor positive – meaning that they respond to oestrogen and oestrogen makes them grow. The problems with tumours in post-menopausal women are that there is not enough estrogen around. Millions of women worldwide continue to get this dreaded diagnosis every year. So what can we do?[11]

With no oestrogen being produced in post-menopausal women, many breast cancer tumors derive a nefarious plan. Cancer cells have the metabolic ability to make their own oestrogen. Seventy percent of breast cancer cell's synthesis oestrogen themselves using an enzyme called aromatase: an enzyme that synthesizes testosterone into oestrogen.

And so drug companies started to produce a number of aromatase inhibitor drugs, which are still being used as chemotherapeutic agents for the treatment towards breast cancer: Tamoxifen, Paclitaxel, Docetaxel and others.[12] But, by the time you are on chemo, it can be too late, as the tumor has already grown or metastasized. Researchers started screening hundreds of natural dietary components in the hope of finding something that actually targets and blocks the function of this enzyme.[13]

Now to do this, you need a lot of human tissue and where are you going to get it from?

Where are you going to get discarded female tissue? They used discarded human placentas. Women actually complied and consented in donating their discarded placentas after giving birth, in-order to further aid this critical-line of research.[14]

After years of searching, researchers found 7 vegetables extracts in vitro that could be a potential dietary source for aromatase inhibitors. Here are the 7 types of vegetables they used in their study:[15]

- Bell peppers,
- Broccoli,
- Carrots, Celery,
- Green onions,
- Spinach and
- White-Button Mushrooms

All 7 different vegetables had reduced aromatase activity by around 20%, except for one important vegetable. They finally found that best fungi, which most people still consider as a vegetable, which actually had the ability to inhibit aromatase activity by up to 65-80%. This food is the "**Common White-Button Mushroom.**"[16]

Important facts: Mushrooms also contain polysaccharides called "beta-glucans" that increase DNA and RNA in the bone marrow where immune cells (like macrophages and T-cells) are made – Beta-glucans enhance immunity to help our body fight against disease.

Researchers at Alpha-Beta Technology in Massachusetts examined the effects of beta-glucans on human blood and found some astonishing facts:

Beta-glucans enhance the growth of myeloid cells and megakaryocytic progenitor cells (which develop into immune cells) and triggered a burst of free radicals in white blood cells that enhances these cells antibacterial activity. More interestingly was they also found that human white-blood cells bacterial-killing capacity was proportional to the beta-glucan dose –the more you eat mushrooms the more you enhance the capability of the immune system to fight off disease, especially bacterial infections.

Mushrooms: Raw or Cooked?

Dr. Joel Fuhrman is a board-certified family physician, and nutritional researcher, who specializes in preventing and reversing disease through nutritional and natural methods.

Dr. Joel Fuhrman recommends to his patients to stop eating raw mushrooms, out of a fear of a natural toxin in regular white mushrooms called Agaratine. Mushrooms have some amazing health benefits, but it's advised to cook them well – in-order to destroy the toxin Agaratine and to get all the mushrooms' nutritional value: as the cell walls of mushrooms cannot be digested unless they are tenderized by heat.

So eat your mushrooms. Approximately 100 grams of "white-button mushrooms is all you need on a daily basis; in-order to help prevent and inhibit breast cancer growth.

Again! Why do Asian Women Have Less Breast Cancer?

Mushrooms clearly show to appear to work in a lab to suppress breast cancer growth, but what about in the real-world where people are involved?[17]

Breast cancer is the most common among women worldwide: with a sixfold variation in incidence between high-risk regions (e.g. Europe and North America) and low-risk regions (e.g. Asia).[18] Maybe it's because of the higher intake of vegetables, green tea and soy consumption within Asian populations? If anything, green tea may only drop breast cancer risk by about a third.[19]

Soya works better, but only it appears if you start young.[20] Soy intake anytime is associated with decreased breast cancer risk. But, the strongest, most consistent effect was for child intake: cutting the risk of later breast cancer by as much as 50 percent. However, if you don't start consuming soy until teens or adulthood, it's only associated with a more green tea type (around 25% drop of breast cancer).[21]

The best is when we consume soy throughout our life, as soy helps reduce the amount of oestrogen in a woman's body, and so, reducing a woman's risk of breast cancer. Soy intake during childhood and adolescence; might provide lifelong protection against breast cancer and sensitize for greater protective effects as an adult. Combined, though, green tea and soy consumption would only account for maybe around a twofold difference in breast cancer risk, and not a sixfold drop.[22]

So the researchers looked into what else Asian women were eating. They already had the intriguing laboratory mushroom data, and so they asked a thousand breast cancer patients how many mushrooms they ate. Then they asked the same questions to a thousand healthy women who they tried to match the cancer patients as similarly as possible, by age, height, body weight, exercise, smoking status, etc. Based on the answers; the researchers calculated that women who averaged at least a certain daily serving size of mushrooms appeared to drop their odds of getting breast cancer by about 64%. What was their average serving size? They concluded, that amazingly, it's been just a lousy half a mushroom per day.[23]

Who eats half a mushroom a day? Well, that was averaged over a month. So, compared to women who didn't regularly eat any mushrooms; those who ate just 15 or more per month, appeared to dramatically lower their risk. Similar protective in the study was found for dried mushrooms and if you combine mushrooms with green tea; sipping a half a teabag worth of green tea per day, along with that half a mushroom, was associated with nearly 90 percent drop in breast cancer odds.

In conclusion: Studies on mushrooms have shown that a daily intake of the common white-button mushroom may have a significant chemo preventive effect with regard to breast carcinogenesis. White-button mushrooms are relatively inexpensive and readily available in market stores, and therefore, are a feasible addition to a healthy dietary cancer regimen.[24]

The Kings & Queens of Anti-Cancerous Plant Foods

If the Cruciferous and Allium family of vegetables are the kings of anticancer foods, then mushrooms, apricot seeds and flax seeds, orange and grapefruit piths, whole apples, berries and lemons may be considered the queens for breast cancer, and other cancers.

Berries are also anti-cancerous fruits, but remember that they are *sweet* fruits, so we do not want to feed the cancer: until the cancer is completed controlled and eradicated.

Fruits That Fight Back Against Cancer Proliferation

What about fruits? Have fruits also fought and target cancer? And if yes, which fruits work best as anti-cancer foods in humans?

A 2002 study, reviewed in the Journal of Agriculture & Food Chemistry,[25] drew up a graph to determine which of the 11 most common fruits eaten in the United States to find out which of them fought against cancer cell proliferation or increased concentrations. The fruits chosen in the study included: pineapples, pears, oranges, peaches, bananas, grapefruit, red grapes, strawberries, apples, lemons and cranberries.

The researchers decided to use human liver cancer cells in this study. Basically, if you drip plain water (water was used as a control) on cancer cells, nothing transpires. Liver cancer cells start out at 100% growth, and they keep powering away at 100% growth.

But, once the researchers dripped the juices of these fruits on liver cancer cells, they found that pineapples, pears and oranges did not seem to do much better. Peaches start to diminish liver cancer growth by about 10%. Bananas and grapefruits were about 4 times better, dropping cancer growth rates by about 40%.

Red grapes, strawberries and apples did even better, cutting cancer cell growth up by 50%, and at only half the dose.

We have all heard the saying: "An apple a day keeps the doctor away". And this quote might actually be very true, as this study clearly shows.

But there were 2 other fruits that really reduced and caused a dramatic drop in liver cancer proliferation at just tiny doses: lemons and cranberries were the clear winners.

So, if you look at the effective dose required to suppress liver cancer cell proliferation, apples are more powerful than bananas, but lemons and cranberries are even more powerful than apples.

Therefore, ladies and gentlemen, eat your red grapes, strawberries, and apples, but, especially, eat your lemons and cranberries. It is a pity that not more fruits were included in the study. It would have been interesting to see how many other Western and tropical Asian fruits, would have done in comparison to the 11 fruits studied in this research.

For example, it would have been useful to test the anti-cancerous power of many other fruits: like blackberries, blueberries, raspberries, cherries, dragon fruit, limes, avocado, jakfruit, star fruit, snake fruit (Salak), noni, goji berries, durian fruit, papaya, mangosteen, kenondong fruit, mango, matua, rambutan, soursop, pulasan, pomegranates, kiwi fruit and many other tropical fruits.

We shall now look at another study on the health benefits of apples, and apple peels against cancer.

A major recent review found that compared to those eating less than an apple a day, those just eating 1 or more have less risk of some cancers: the the mouth, the larynx, the breasts, the colon, the kidney and the ovaries.[26]

The higher the apple concentration, the higher the rate of dropping human cancer cells. The same was noticed with breast cancer. Apples appeared to work best against oestrogen receptor negative breast cancers, which are much harder to treat, than oestrogen receptor positive breast cancer. How do apples do that?

Apples Actually Block Tumour-Cell Formation

Carcinogens cause DNA mutations, and then oxidation. Inflammation, toxins, and hormones cause it to grow and, finally, to metastasize. Which steps do apples block? It is amazing, but they block *all* of these steps.

What About Apple Peels?

Most people remove the apple peel before consuming apples mainly because of the amounts of pesticides apple peels may contain. In the last few years, half a dozen studies have publicized the benefits of apple peel extracts against cancer cells.[27]

We already know that the more apples we eat, the lower our risk of several cancers. The peels of apple and other fruits, which have been shown to process exceptionally high concentrations of antioxidants, are often discarded.

In a recent study,[28] apple peels were placed in a blender to extract the juice and then dripped on breast and prostate cancer cells.

This is not chemotherapy, but just apple peel extract. The mechanism involved was this: apple peels have an anti-proliferative effect on the tumour suppressor protein, maspin, in prostate and breast cancer cells. Maspin is a type II tumour suppressor gene that has been shown to have tumour suppressor, anti-angiogenic, and anti-metastatic properties in both prostate and breast cancer cells.

The tumour cells found a way to turn this tumour suppressor gene off, and apple peels apparently turned it back on.

This could be due to the up-regulation of this tumour suppressor gene as more and more apple peel extract is added to each of the cancer types. It was concluded that apple peels may process strong anti-proliferative effects against cancer cells, and so, they should not be discarded from our diet.

I would recommend that you wash your apples thoroughly, in-order to remove excess pesticide residues, and still eat your apple peels to help the body fight back against malignant cells.

Anti-Angiogenesis:
Cutting-Off Tumour Blood Supply with Specific Plant-Foods

Recent studies have estimated that about one-third of the most common cancers in high-income countries can be prevented: by consuming a healthy plant-based diet, being physically active, and maintaining a healthy weight.[29]

One of the extraordinary ways that plants can help in reducing tumour promotion and especially tumour progression, is by cutting off blood supply lines to cancer – which is a medical metabolic process called "anti-angiogenesis".

A tumour cannot grow without a blood supply. Currently, it is believed that without a proper blood supply a tumour mass cannot exist in a volume greater than about the size of the ball at the tip of a ball-point pen. This indicates that angiogenesis (*angio* means vessel; *genesis* refers to the creation of new vessels) is fundamentally critical to tumour growth.[30]

Each one of us has cancer cells in our bodies right now. For example, by age 70, microscopic cancers are detected in the thyroids of virtually everyone. Most of these tumours at no time do not become clinically significant; nor do they cause problems. This leads to the concept of

Foods that Block Cancer Angiogenesis

- Garlic
- Onions
- Leeks/shallots
- Kale & bok Choy
- Collard greens/spinach
- Brussels sprouts
- Broccoli
- Mushrooms
- Artichokes
- Cabbage
- Red cabbage
- Peppers
- Citrus fruits
- All types of berries
- Apples/apple peels
- Pineapples
- Red grapes & red wine
- Pumpkin
- Tomatoes
- Lavender/ginger
- Turmeric
- Oranges inside white and pale (pith)
- Grapefruit inside white (pith)
- Pomegranates
- Vitamin B17 (laetrile)
- Selenium

"cancer without disease" as a normal state during aging. Cancer cells are commonly present in the human body, but they cannot grow into tumours larger than microscopic dot size, and no more than 10 million cancer cells before needing to get hooked up to a blood supply.[31]

Therefore, tumours have a unique ability to release angiogenic factors (AF) – chemicals that cause new blood vessels to sprout into the tumour. The most important AF is called "Vascular Endothelial Growth Factor" (VEGF). However, we can suppress VEGF with VEGGIES and other plant-based foods!

Many difference plant foods can actually block the stimulation of new blood vessels to tumours. They are ideal for prophylactic long-term use for breast cancer: because of their reliability, availability, safety, and affordable price. Dietary agents used to suppress angiogenesis can be an important step in the prevention, treatment and management of breast cancer and, in fact, all types of cancers.

The extensive ranges of studies provide convincing evidence that dietary phyto-constituents possess the unique ability to affect tumour angiogenesis, which may be deemed advantageous in the prevention and treatment of human breast cancer.[32]

The problem is that the majority of these studies have mostly been performed in a Petri dish; where one stimulates human blood vessel cells that start forming blood vessel-like tubular structures, trying to make new capillaries (small blood vessels) to feed the tumour.

Where do researchers get new blood vessels from? The answer is: human umbilical vein endothelial cells (HUVECs). They get them from discarded umbilical cords or, more controversially, from the eyes of aborted foetuses. Either way, you can stimulate blood vessel formation with the tumour compound VEGF, and then abolish that effect with plant compounds.[33]

However, if you add plant phytonutrients, like polyphenols, flavones (apigen), glucosinolates, carotenoids, isothiocyanates, indoles, lycopene, luteolin, etc., these phytochemicals help block these tubes (blood capillaries) like formations.

Fisetin, a natural flavonol, a phytonutrient found commonly in strawberries and other fruits and vegetables, has been shown to possess efficacy against many cancers,[34] and it just shrinks the formation of new blood vessels right the way down.

In a 2003 article, M.F. McCarty wrote that:

> The daily consumption of natural plant foods containing adequate flavonoids and other anti-angiogenic phytochemicals could be beneficial for the prevention of cancer metastasis (spread) and, in so doing, improves cancer prognosis.[35]

Given the power of plants, one might speculate that the foundation of an anticancer approach would be to "eat a very low fat, whole-food, vegan diet emphasizing lower-glycemic-index starchy foods, and exercising regularly".[36]

Notes for chapter 29

1 Biovin, D., *et al* (2009). Antiproliferation and antioxidant activities of common vegetables: a comparative study. *Journal of Food chemistry*, **112** (2), p.374-380.

2 Greger, M. (September 30th 2013). *Breast Cancer Survival Vegetable. Simple changes in diet and lifestyle may quadruple a woman's survival rate from breast cancer*, **15** video 1. Retrieved from http://nutritionfacts.org/video/breast-cancer-survival-vegetable/ . Greger, M (November 9th 2009). #1 *Anticancer Vegetable*, **3** video. Retrieved from http://nutritionfacts.org/video/1-anticancer-vegetable/

3 Biovin, "Activities of common vegetables," 2009.

4 Hans, Clevers. (2011). Focus on cancer: The cancer stem cell: Promises, Promises and Challenge. *Natural Medicine*, pp.313-319. doi:10.1038/nm.2304

5 Greger, M. (March 12th, 2012). *Broccoli vs. Breast Cancer Stem Cells: A new theory of cancer biology - cancer stem cells – and the role played by sulforaphane, a phytonutrient produced by cruciferous vegetables*, **8** video. Retrieved http://nutritionfacts.org/video/broccoli-versus-breast-cancer-stem-cells/

6 Ibid.

7 Yanyan li, *et al* (2010). Sulforaphane, a dietary component in broccoli/broccoli sprouts, inhibits breast cancer stem cells. *Clinical Cancer Research*, **16**, pp.2580-2590.

8 Ibid.

9 Cornblatt, S. Brian *et al* (2007). Preclinical and Clinical evaluation for sulforaphane for chemoprevention in the breast. *Carcinogenesis*, **28**(7), pp.1485-1490.

10 Ibid.

11 Adams, *et al*. (2009). Phytochemicals for Breast Cancer Prevention by Targeting Aromatase. *Department of Surgical Research, Beckman Research Institute of the City of Hope. Front Bioscience*, **1**(14), pp-3846-63. Ibid, Greger, M (2011).

12 Lonning P.E. (2004). Aromatase Inhibitors in Breast Cancer. *Endocrine-Related Cancer*, **11**, pp. 179-189

13 Greger M. (2011). *Vegetable's vs Breast Cancer: Mushrooms may help prevent breast cancer by acting as an aromatase inhibitor to block breast tumor estrogen production.* Volume **5** video. Retrieved from http://nutritionfacts.org/video/vegetables-versus-breast-cancer/

14 Adams, "Phytochemcials for Breast Cancer," 2009.

15 Ibid.

16 Ibid.

17 Baiba JG, *et al.*(2001). White button mushroom phytochemicals inhibit aromatase activity and breast cancer cell proliferation. *Journal of Nutrition*, **131**(12), pp.3288-3293. Greger M. (December 27th, 2012). *Why do Asian Women have Less Breast Cancer?* Volume **17** Video. Retreived from http://nutritionfacts.org/video/why-do-asian-women-have-less-breast-cancer/

18 Parkin DM, *et al.*(2006). Use of statistics to assess the global burden of breast cancer. *Journal of Breast*, **12** (Suppl 1), pp.S70-80.

19 Anna H. Wu, Lesley M. Butler.(2011). Green tea and Breast Cancer. *Molecular Nutrition & Food research*, **55**(6), pp.921-930.

20 Korde LA, *et al.* (2009). Childhood soy intake and breast cancer risk in Asian American women. *Cancer Epidemiological Biomarkers Preview*, **18**(4), pp.1050-1059

21 Anna H. Wu, *et al.*(1998). Soy intake and risk of breast cancer in Asians and Asian Americans. *American Journal of Clinical Nutrition*, **68** (suppl), pp.1437S–43S.

22 Ibid.

23 Min Zhang, *et al.* (2009).Dietary intake of mushrooms and green tea combine to reduce the risk of breast cancer in Chinese women. *International Journal of Cancer*, **124**(6): pp-1404-1408. Greger, *"Why do Asian Women have Less Breast Cancer?"*, 2012.

24 Chen S, *et al.* (2006). Anti-aromatase activity of phytochemicals in white-button mushrooms (Agaricus bisporus). *Cancer Research*, **66**(24), pp.12026-34. Greger M.(2011). *Breast Cancer prevention. Which Mushroom is Best?* Volume **5** video. Retreived from http://nutritionfacts.org/video/breast-cancer-prevention-which-mushroom-is-best/

25 Rui Hai Liu, *et al.* (2002). Antioxidant and Antiproliferative Activities of Common Fruits. *Journal of Agriculture & Food Chemistry*, **50**(25), pp.7449-7454.

26 Gerhauser, C. (2008). Cancer Chemopreventive Potential of Apples, Apple Juice, and Apple Components, Affiliation. *Division of Toxicology and Cancer Risk Factors. German Cancer Research Centre (DKFZ), Heidelberg.*

27 Wolfe, K., Wu, X.Z., & Liu, R.H. (2003). Antioxidant activity of apple peels. *Journal of Agriculture and Food Chemistry*, **51**(3), pp.609-614. Reagan-Shaw, S *et al.* (2010). Antiproliferative effects of apple peel extract against cancer cells. *Department of Dermatology, University of Wisconsin, USA. Nutrition and Cancer*, **62**(4), pp. 517-524.

28 Ibid.

29 Miller, P.E., *et al.* (2012). Phytochemicals and Cancer Risk: A Review of the Epidemiological Evidence. *Nutrition of Clinical Practise*, **27**(5), pp.599-612. Greger, M. (2013). *Anti-angiogenesis: Cutting off tumour blood supply with plant foods*, Volume **13**. Retrieved

from http://nutritionfacts.org/video/anti-angiogenesis-cutting-off-tumor-supply-lines/

30. Reuben, S.C., *et al.* (2012). Modulation of angiogenesis by dietary phytoconstituents in the prevention and intervention of breast cancer. *Molecular Nutrition Food Research*, **56**(1), pp.14-29. Li, W.W *et al.*(2012). Tumour Angiogenesis as a Target for Dietary Cancer Prevention. *Journal of Oncology*, Article ID 879623.

31. Greger, "Anti-angiogenesis," 2013.

32. Reuben, "Modulation of angiogenesis," 2012.

33. Ibid.

34. Bhat, T.A., *et al.*(2012). Fisetin inhibits various attributes of angiogenesis in vitro and in vivo-implications for angioprevention. *Carcinogenesis*, **33**(2), pp.385-393.

35. McCarty, M. F. (2003). A wholly nutritional 'multifocal angiostatic therapy' for control of disseminated cancer. *Medical Hypothesis*, **61**(1). pp.1-15.

36. Greger, "Anti-angiogenesis," 2013.

30

How Stress Influences Human Health

If you ask what is the single most important key to longevity, I would have to say it is avoiding worry, stress and tension. And if you didn't ask me, I'd still have to say it.

The late George Burns (1896 -1996)
An American comedian, award-winning actor and best-selling writer

The Physiological/Psychological Mechanisms, and Effects of Stress

Following the perception of an acute stressful event, there is a cascade of changes: in the nervous, cardiovascular, endocrine, and immune systems. These changes constitute the stress response and are generally adaptive, at least in the short term.[1]

Two features in particular make the stress response adaptive.

1. Stress hormones are released to make energy stores available for the body's immediate use.
2. A new pattern of energy distribution emerges. Energy is diverted to the tissues that become more active during stress, primarily the skeletal muscles and the brain. Cells of the immune system are also activated and migrate to "battle stations"

Simply put, during times of acute crisis, eating, growth, and sexual activity may be a detriment to physical integrity and even survival.[2]

Stress hormones are produced by the somatic nervous system (SNS) and hypothalamic-pituitary adrenocortical axis. The Autonomic Nervous System stimulates the adrenal medulla to produce catecholamines (e.g., norepinephrine, epinephrine and dopamine). In parallel, the paraventricular nucleus of the hypothalamus produces corticotropin releasing factor, which in turn stimulates the pituitary to produce adrenocorticotropin.

Adrenocorticotropin then stimulates the adrenal cortex to secrete cortisol. Together, catecholamines and cortisol increase available sources of energy by promoting lipolysis (the breakdown of fats into usable sources of energy, like fatty acids and glycerol) and the conversion of glycogen into glucose.[3]

The acute stress response can become maladaptive, if it is repeatedly or continuously activated in long-term chronic stress situations.[4] Why is this so?

1) Chronic SNS stimulation of the cardiovascular system due to stress leads to sustained increases in blood pressure and vascular hypertrophy.[5] Muscles constrict the vasculature

thicken, producing elevated resting blood pressure and response stereotypy, or a tendency to respond to all types of stressors with a vascular response.

2) Chronically elevated blood pressure forces the heart to work harder, which leads to hypertrophy of the left ventricle. Over time, the chronically elevated and rapidly shifting levels of blood pressure can lead to damaged arteries and plaque formation.[6]

3) The elevated basal levels of stress hormones associated with chronic stress also suppress immunity by directly effecting cytokine profiles.[7] Specific immune cytokines are molecules that are essentially needed for the immune system to function adequately to combat the fight against inflammatory responses, microorganisms and other diseases.

4) During periods of chronic stress, in the otherwise healthy individual, cortisol eventually suppresses proinflammatory cytokine production. Prolonged proinflammatory cytokine production may also adversely affect mental health in vulnerable individuals. During times of illness (e.g., the flu), proinflammatory cytokines feed back to the CNS and produce symptoms of fatigue, malaise, diminished appetite, and listlessness, which are symptoms usually associated with depression. It was once thought that these symptoms were directly caused by infectious pathogens, but more recently, it has become clear that proinflammatory cytokines are both sufficient and necessary (i.e., even absent infection or fever) to generate sickness behaviour.[8]

Studies have shown that psychological stress affects the immune system, and in that way, that is one mechanism by which cancers are able to progress. When you have cancer, you truly have to step back and look at all aspects of your life: not only, your diet, because you seriously have to ask yourself "Is the way I'm living killing me, because it may just well be?" And if you have cancer, it probably is. It's the way you are living; it's your whole life, not just your diet. Obviously, your lifestyle has to change and your stress levels have to be controlled.

Host Factors and Cancer Progression

Evidence on the role of psychological disturbances in cancer initiation has been equivocal; support continues to grow for links between factors: like stress, depression, social isolation, and progression of cancer.

Clinical studies show that stress-related processes can impact pathways – implicated in cancer progression, including immuno-regulation, angiogenesis and invasion.

The stress response results in activation of the autonomic nervous system and the hypothalamic-pituitary-adrenal axis. Factors released from these pathways have been shown to have a direct effect on the tumour micro-environment: resulting in a favorable environment for tumour growth and progression.[9]

Stress: Factors Released and Their Effects on the Tumour Micro-Environment

Immune and Stromal Cells	Cancer cells
↓ Natural Killer cells (NK) function	↑ Migration and invasion
↓ T-Cell activity	↑ The production of angiogenic factors (vascular endothelial growth factor (VEGF), interleukin -6 (IL-6) and signal transducer and activator of transcription factor-3 (STAT 3).
↑ Regulatory T-cells	
↑ Tumour-associated macrophages (TAM) and (matrix metalloprotinease (MMP) and inflammatory cytokines	

Both microenvironment pathways influence cancer growth and progression and the quality of life to cancer patients.

Furthermore, there is evidence that stressful life events are cause for the onset of depression.[10] Stressful life events often precede anxiety disorders as well.[11] Psychosocial interventions have proven useful for treating stress-related disorders, and may influence the course of chronic diseases.

You have to transition your whole life. A lifestyle change is desperately needed. Thinking and feeling favorably and positively, psychologically and physically, without any doubt is a must. You have to say to yourself, "I can beat this disease." The biggest challenge for any medical professional is "Patient Compliance." If the patient does not genuinely believe that your choice of treatment cannot aid them, then it may be a lost cause.

The sick patient needs to complete the course, and the therapist needs to convince patient's that your treatment will succeed in its brilliance in eradicating cancer. Cancer is a lifestyle battle. Preventing or reversing cancer involved – choosing to make good decisions, choosing to stick to the cancer management program, choosing hard-core nutrition over junk foods, choosing to change your life and get rid of all your stresses, whilst creating a healthy and peaceful life, full of confidence and self determination, that is conducive to healing.

Psychosocial interventions, such as cognitive-behavioral stress management (CBSM), have a positive effect on the quality of life of patients with chronic disease. Such interventions decrease perceived stress and negative mood (e.g., depression), improve perceived social support, facilitate problem-focused coping, and change cognitive appraisals, as well as decrease SNS arousal and the release of cortisol from the adrenal cortex.[12]

Psychosocial interventions also appear to help chronic pain patients reduce their distress and perceived pain, as well as increase their physical activity and ability to return to work.[13]

To achieve these goals and help the patient cope with stressful situations (death of a loved one, ill health and other stressful events), one must have an excellent cancer medical management team, friends and family support, that can assist the patient whilst undergoing treatment. Persons that correctly understand and truthfully acknowledge the importance of dietary and lifestyle changes for disease reversal. If this is the case, then stress management, diet and lifestyle interventions may accomplish these tasks. Overall, lifestyle medicine will aid the ill patient, in giving them back their self-confidence and hope in believing that they *can* heal themselves, mentally and physically from disease.

Notes for chapter 30

1. Selye, H (March 1st, 1978). *The Stress of Life* (2nd Ed). Published by McGraw-Hill, New York.
2. Scott, D *et al.* (2005). STRESS AND HEALTH: Psychological, Behavioural, and Biological Determinants. *NIH Public Access: Annu Rev Clin Psychol*, **1**, pp. 607–628.
3. Brindley DN, Rolland Y *et al.* (Nov, 1989). Possible connections between stress, diabetes, obesity, hypertension and altered lipoprotein metabolism that may result in atherosclerosis. *Clinical Science*, **77**(5), pp.453-61.
4. Ibid.
5. Henry JP, Stephens PM, Santisteban GA. (Jan, 1975). A model of psychosocial hypertension showing reversibility and progression of cardiovascular complications. *Circ Res*, **36**(1), pp.156-164.
6. Brownley KA, Hurwitz BE, Schneiderman N. (2000). Cardiovascular psychophysiology: Function methodology, and use in pathophysiological investigation. In Cacioppo, JT, Tassinary LG, Berntson GG (Eds.), Handbook of Psychophysiology (pp.224-264), Cambridge University Press, New York.
7. Roitt I, Brostoff J, Male D.(1998). *Immunology* (5th ed), p.125. London, Mosby Int.
8. Larson SJ, Dunn AJ. (Dec, 2001). Behavioural effects of cytokines. *Brain Behav Immun*, **15**(4), pp.371-387.
9. Lutgendorf, S.K *et al.* (2010). Host factors and cancer progression: biobehavioral signaling pathways and interventions. *J Clin Oncol*, **28**(26), pp.4094-4099. doi: 10.1200/JCO.2009.26.9357.
10. Kendler KS, Karkowski LM, Prescott CA *et al.*(1999). Causal relationship between stressful life events and the onset of major depression. *American Journal of Psychiatry*, **156**(6), pp.837-841.
11. Faravelli C, Pallanti S *et al.* (May, 1989). Recent life events and panic disorder. *Am J Psychiatry*, **146**(5), pp.622-6.
12. Schneiderman N, Antoni MH, Saab PG, Ironson G. (2001). Health psychology: psychosocial and biobehavioral aspects of chronic disease management. *Annu Rev Psychol*, **52**, pp.555-580.
13. Morley S, Eccleston C, Williams A. (March, 1999). Systematic review and meta-analysis of randomized controlled trials of cognitive behaviour therapy and behaviour therapy for chronic pain in adults, excluding headache. *Journal of Pain*, **80**(1-2), pp.1-13.

31

Transition to a Plant-Based Diet and the Significance of Detoxification

We should all be eating fruits and vegetables as if our lives depended on it – because they do.

Michael Greger, M.D.
A physician, author, and internationally recognized speaker on nutrition, food safety, and public health issues

Throughout this book, I have provided you enough medical evidence and I have also very hard-worthily spent thousands of hard-working hours trying to found the scientific proof of the health benefits of a plant-based diet and lifestyle interventions: for disease prevention and disease reversal. However, as much as it is hard to stop smoking or stop consuming sugary processed foods, meat and dairy products, it is just as hard to change your overall diet & lifestyle.

So the key is to try to start *reducing* animal-based and refined food products; in order to gradually eliminate them from your daily dietary regime. The first step to lifestyle medicine is diet: hopefully and perspectively the other intervention strategies correlating and dealing with a healthy diet and lifestyle, may instantly follow.

Some people transit straightaway to a plant-based diet – owing to improving human health, environmental and moral ethical animal welfare reasons.

We should not start to change our eating habits and lifestyle, when we already have a chronic disease. Simply put, it is much harder to cure any form of disease once the disease has been given the time to progress. We need to start today. Today is never too late.

My personal option (and the way it worked for me) is that you should be strong and determined *not* buy and consume any types of foods that are detrimental to the body: overall, this will help you to choose, buy, eat and add healthier plant foods to your overall dietary regime.

Once you experience the benefits these changes in dietary habits will offer to your overall health, you will automatically appreciate and enjoy eating foods that will automatically make you feel physically and mentally healthier.

I have been strongly influenced by many clinical physicians who understood the real health benefits of a plant-based diet: that truly and bravely spoke out to encourage a healthier diet and lifestyle. I watched videos and studied books: about diet and cancer, diet and heart disease, and other informative health books and videos that clearly focused on diet and lifestyle changes; emphasizing on how a dietary and lifestyle plan could prevent and even reverse patients suffering from diabetes, heart disease, many forms of cancer and other related-degenerative diseases.

I was determined to change my own eating habits, because I knew the long-term health benefits this could bring to my own health. And it has. I feel stronger, cleaner, more alive and aware of things around me. My mind is more open and my own blood pressure, blood glucose and other clinical profiles are normal. I also lost weight naturally, without taking drugs or other medications – via changing my eating habits, including eliminating animal foods and introducing more plant foods to my over-all daily-dietary regime. I would surely recommend that you combine a healthy diet with moderate amounts of exercise.

Walking or rebounding (mentioned in an earlier chapter) for 20-30 minutes, 3-4 times a week, is one of the better ways to exercise: to help remove the build-up of free radicals, eliminating toxins from the body and overall boosting the function of the human immune system.

I personally would recommend that you eat food two hours before exercising. Consume foods that contain high amounts of antioxidants in the form of natural juices, such as – broccoli, spinach, beets, lemons, oranges, apples, berries, and grounded flaxseeds. During exercise toxins can build up in our body, so extra consumption of foods that contain antioxidants can help flush out these harmful-metabolic toxins which are produced during exercising.

There are also hundreds of vegetarian and vegan recipe cook books to choose from which can be brought or obtained from shops and libraries or downloaded from the world-wide web. These will start you off on the right path, by giving you knowledgeable ideas about how to use a variety of plant foods in making many healthier food variations and delicious vegan recipes, to help improve your own and your family's daily-dietary eating habits and overall health.

You will be amazed how easy it can be to make your own breakfast, lunch and dinner, by just mixing and eating vegetables, fruits, nut, seeds, legumes and lentils, topped up with or added to whole grains with small amounts of healthy herbs and spices. You do not need animal foods to make your food taste good and look appealing.

Even for workaholics, busy families with children, persons who do not have the time to prepare a meal, there are alternative cooking methods and food equipment: to help you save time and money.

Eating large amounts of fruits and vegetables can be time-consuming, and not many people today have the time for it. The best way to save time is to buy food equipment that will help you prepare these food products for the morning, afternoon or even for a quick refreshing drink when you return home after a hard-working day.

A good blender and juicer would be the solution to this problem. By juicing your vegetables and fruits it will provide you with most of the nutrient benefits you need to help boost your energy for the whole day, especially for those who do not have the time to cook or may not feel like taking an hour or more to cook a healthy meal. Juicing allows the nutritional properties to be more available and even more accessible and, at the same time, increase your body's rate of absorption. Nutrients are absorbed more efficiently by the bloodstream, thus alleviating the digestive tract of a prolonged systemic function. More importantly, vegetable and fruit juicing can induce major health prevention processes. Phytochemical synthesis of antioxidant properties against free radicals is a vital source of detoxification mechanisms, that also promotes longevity and good health.

But juicing does have one major downside: the elimination of dietary roughage. Nevertheless, plant fibre mass can be obtained in other raw food preparation and cooking.

So, please, remember to try to *eat and juice* your fruits and vegetables. You can make wonderful soups and salads with whole vegetables, topped up sliced fruits, nuts and grounded seeds: these will retain most of the natural nutrients and fibre content. So, adding vegetable salads, soups and juices should be a must in your daily dietary regime.

I would highly recommend juicing at least once a day: in the morning or a couple of hours before you go to bed. Unfortunately, man-made industrial toxins have polluted our environment and the foods we eat and so, we all need to shield ourselves all day long from these harmful pollutants. For cancer patients more frequent juicing is a must. Juicing should be used in any cancer management treatment. Eat plant foods and you will at least get the phytochemicals and their antioxidant properties to help mop-up these harmful compounds. Animal products DO NOT have these compounds: except in very minute quantities.

Try to mix green vegetables with carrots, beets and apples. A good juice for cleansing the liver is a combination of beetroot and apple juice. There are many books to help you choose, which plant juices taste the best when combined with fruits or vegetables or with both.

I have tried my best to consume as many nutrients and phytochemicals in my daily diet as possible. This will help boost my liver in phase 1 and 2 of the detoxification process. It will also protect against and reduce free radical damage caused by everyday normal metabolism and by the inhalation and metabolism of harmful industrial pollutants. These healthy strategies will reduce the risk of contracting any acute or long-term chronic degenerative disease. You will feel more content, safe, relieved and physically and mentally stable.

Once you transition to a plant-based diet you may feel some unpleasant side-effects: headaches, stomach discomforts, fatigue and nausea. These are only mild symptoms of the body's healing process. Toxins are being eliminated from the body, and so some people may develop some side-effects for the first 2 to 3 weeks. Once the body has been cleansed these symptoms will gradually disappear. Do not think that eating healthier plant-foods is not compatible with your body's new lifestyle. These are just normal symptoms in-which the human body is showing you that it is healing from all the toxic build-up you have accumulated from your entire life, since your birth.

Detoxification is Essential to Human Health

Why is detoxification of the body an important step in healing the human body? Detoxification is an essential process because it eliminates excess toxins from the body by neutralizing or transforming them, and clears excess mucus and congestion. A poor diet, poor digestion, a sluggish colon, reduced function of the liver, and dismal elimination from the kidneys: all lead to increased toxicity, and a lack of oxygen at the cellular level.[1]

Refined sugars and lack of oxygen may create the perfect environment for anaerobic microbes (bacteria, parasites, viruses and fungi) to rapidly propagate. Our cells can then literally become infected with these microbes and eventually cause our cells to either die or "morph" into cancer cells.

Once our body (specifically the liver, kidneys, gallbladder and bowels) loses the ability to process all the toxins and pollutants, the oxygen supply will lessen, the immune system begins to collapse, the body's pH becomes more acidic (acidosis), and we will created a perfect breeding ground for deadly microbes, parasites and cancer.

These microbes are the end-result when our body's immune system has lost its ability to protect its cells from carcinogens. These invaders may act as the actual catalyst for cancer and nearly all other diseases, hijacking a healthy aerobic cell. The bacterium and virus invaders will start exhausting the cell's oxygen and energy supply, until the cell either dies or mutates into an anaerobic (i.e. cancer) cell.

Now this anaerobic cell relies on fermenting sugar to produce energy. The battle of cancer is truly fought at the cellular level in an effort to cleanse the body of microscopic invaders; while radically changing the body's internal terrain back to a healthy homeostatic state. This is why cleansing is so vital to all of us.

Cancer cells prefer an oxygen-deleted environment, an acidic blood pH, and the usage and fermentation of glucose to produce lactic acid in order to continue creating such environment and initiate growth, manifest, and spread to other organ tissues. Basically, the "right" conditions for the development of cancer environment have been achieved.

It wisely makes sense first to remove these harmful toxins from our bodies. We need to detoxify to heal the body and then treat the cancer properly. So it is extremely important for cancer patients, and also for non-cancer patients, because our bodies are actually toxic-waste dumps.

I am still amazed that people think it is normal to have only 1 or 2 bowel movements per week. You should be having 1 or 2 bowel movements per day, *every* day.

More than just symptoms of constipation, infrequent bowel movement can mean a toxic overload that can play havoc to your body. Signs of an imbalanced colon include: bloating, excess gas, irritable bowel, fatigue, haemorrhoids, headache, and backaches.[2]

Have you ever had a headache, a running nose, back pain, constipation, indigestion, mood changes or fatigue? Have you ever wondered what could have caused these symptoms? These could be the symptoms of toxic overload, caused mainly by the air you breathe, the foods you eat, the water you drink, and other normal metabolic body processes.

Unfortunately the environment we live in is a rather toxic place. According to the Centre for Disease Control and Prevention (CDCP) most people are regularly exposed to around 50,000 chemicals, and 2 billion people may be suffering from parasitic infections, 150 million of which in the United States alone.

Even though we might think we are taking proper precautions, it is still hard to avoid all the contaminants. The human body was designed to get rid of these harmful pollutants, but even an efficient body can become overwhelmed by the sheer volume of preservatives, chemicals, pollutants and parasitic contaminants that are absorbed during the day. In fact, the average person may carry around 5lb to 25lb of toxic matter!

These are some of the everyday chemicals or compounds that can be carcinogenic and cause toxic-overload. This is how we are slowly being poisoned from day to day:

1. Bad food choices: Meat and dairy products, refined and processed foods.
2. Body care products: shampoos, soaps, antiperspirants or deodorants, toothpastes and beauty cosmetics.
3. Home-cleaning products: dish-washing soaps, laundry detergents, bathroom and kitchen cleaning products.

4. Fluoridating public drinking water with industrial waste. (Fluoride is known to be a highly toxic chemical that affects the brain, and it should have no place in our food, drinking water and dental care products – contrary to what we have been told by dentists and doctors since we were kids.) Fluoride free toothpaste should always been used.

5. Sunlight: sunlight is extremely important to our health; only excessive exposure to the sun damages the skin and can lead to skin cancer. Yet, humans need some sunshine exposure for the synthesis of vitamin D. We are told to use sunscreens that are supposed to protect us from harmful UV rays, but many sun-blockers are loaded with cancer-causing chemicals, that can be absorbed through the skin.

6. Food preservatives and additives, some of which are carcinogens. Artificial sweeteners, like aspartame and monosodium glutamate are highly toxic, even though it is "legal" to use them and are actually found in most refined foods.

7. Pesticide-laden fruit and vegetables. Even if you eat the world's healthiest foods, they are still contaminated with carcinogens.

8. Vaccines. These contain mercury and formaldehyde, besides other toxic chemicals.

9. Carcinogens (chemical, physical and biological) and environmental factors, such as industrial pollutants like dioxins, polycyclic aromatic hydrocarbons, nuclear and ionizing radiation, x-rays, synthetic chemicals, heavy metals, PCBs and PAHs, etc.

The above factors evidently lead to a suppressed immune system. People get sick and go to their doctors. Instead of doctors explaining the importance of diet and lifestyle changes for disease prevention, most doctors only prescribe pharmaceutical drugs and vaccines, which continue the toxic-chemical onslaught.

The build-up of toxic matter needs to be cleansed from our systems, because every tissue and cell of the body is fed through our bloodstream, which is supplied from the intestines. If our intestines are overloaded with toxins, then so is our blood. The increased exposure to toxins, may, over time, cause havoc to human organs, tissues and cell's, and the whole body, including the digestive system. The deterioration of the digestive system can occur silently over many years, producing no visible or detectable symptoms except very minor, non-specific ones: like fatigue, headaches, back pain, flatulence, bloating, constipation and indigestion. These can be the prelude to the long-term toxic overload that could lead to many chronic problems: obesity, diabetes, peptic ulcers, prostate disease and colorectal cancer, among many other diseases.

In fact it is now believed that 60 to 80 percent of all known cancers are the direct result of the chemicals that are found in our homes, in the air, water and foods, and in other industrial environmental pollutants. Unfortunately, many of us wait until they experience serious health problems to even consider a healthy detoxification digestive system regime.

The initiation steps to detoxify the human colon, liver and kidneys should include:

- Daily exercise
- Drinking plenty of water and apple/lemon juice (2 litres per day)
- Eating a variety of plant foods: fruits, vegetables, nuts, seeds and legumes
- Performing an enema cleanse
- *Not* eating any processed or refined foods or foods of animal origin
  (After completing the detoxification programme) is it vital to continue with a healthy dietary regime.

Personally I highly recommend that you perform a detoxification programme to cleanse your body. Detoxification should be performed at least twice a year to flush out the build-up of these toxins, so that the body may be cleansed. Exercise, drink plenty of water, and include plenty of fresh vegetables, fruit juices, vegetable soups and salads, topped up with nuts and ground seeds in your detoxifying-dietary regime for a minimum of 5-7 days. This treatment should be adequate to cleanse the body from toxins. Coffee enemas or other types of enemas can be performed, but only under the guidance of a qualified physician.

In this context, Dr. Michael Greger's comment is very pertinent:

> Diet and lifestyle changes can reverse our number one killer (heart disease) and other chronic diseases. This lack of awareness results in millions upon millions of totally unnecessary deaths. There are now accepted, basic, nutritional principles that have not changed in decades, yet, people are still unaware of the power of nutrition to improve their lives.[3]

I have been inspired by many wonderful persons who have dedicated their lives in helping others: to take control of their own lives and to understand the power of nutrition excellence and lifestyle changes has towards healing and strengthening human health.

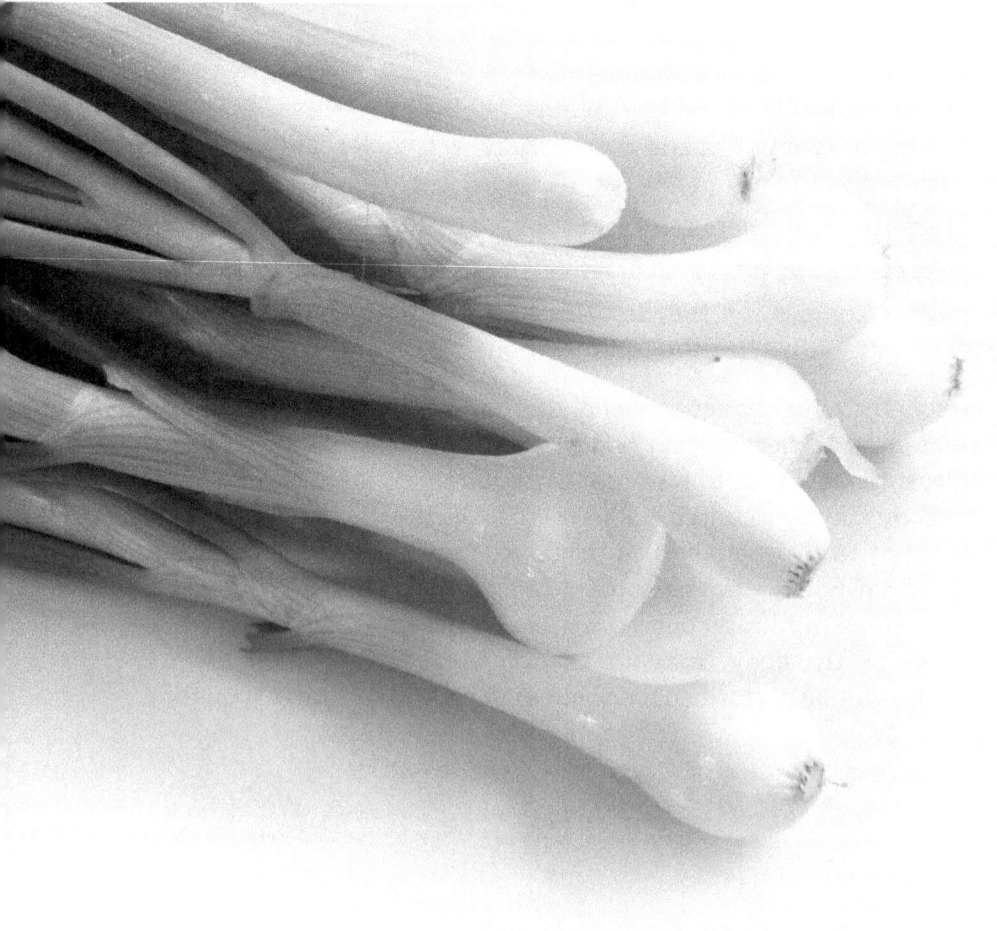

Eating in Moderation: Good or Bad for you?

Often I engage in discussions about the importance of avoiding foods that are high in saturated fats, cholesterol, refined sugars, and foods that are heavy in preservatives, additives and other harmful chemicals. The answer I very often get is: "Isn't it OK to eat everything as long as it's in moderation?" This seems reasonable: most people prefer moderation and are generally wary of extremism. But when it comes to our health, is moderation enough for most people? Let me tell you this.

Is eating in moderation enough to stop the process of disease reversal? If you cut back considerably on fat and cholesterol, should you not be all right? Surely, just a little bit would not hurt?! Wrong! Moderation can kill! And to fully understand why, you have to understand something about metabolism and biochemistry.

Every segment of our bodies is composed of cells. The trillions of cells within the human body are protected by an outer coat called "a cell membrane." This cell membrane is almost unimaginably tiny and delicate, in fact just one hundred-thousandth of a millimetre thick. Yet it is absolutely essential for the integrity and healthy functioning of the cell, as all cells are extremely vulnerable to injury!

Any foods containing saturated fats and excess cholesterol that come from animal fats and plant or animal oils (including dairy foods) initiate an assault on these membranes and, therefore, on the cells they protect. Food high in fats and oils can produce a cascade of free radicals in our bodies – that are especially harmful chemical substances that induce metabolic injuries from which there can be only partial recovery. Year after year, these effects can accumulate and, eventually, the cumulative cell injury is big enough to become obvious, to express itself as, what physicians define a disease.

Plants and grains do not induce the deadly cascade of free radicals. Even better, in fact, they carry an antidote. Unlike animal products and oils, they contain antioxidants that neutralize or mop-up the free radicals and provide effective protection against cancers.[4]

Every time we eat a typical westernized diet we damage the endothelium itself: a thin layer of flat epithelial cells that lines serous cavities, lymph vessels, blood vessels and vessels of the heart. The endothelium plays an important role in maintaining a healthy blood flow. Remember, healthy endothelium cells also produce nitric oxide, which is critical to preserving the health of all blood vessels. Nitric oxide is a vasodilator, that is, it causes the vessels to dilate or enlarge. When there is abundant nitric oxide in the bloodstream, it keeps blood flowing as if the vessels' surfaces were coated or lubricated with Vaseline – eliminating the stickiness of vessels and blood cells that is caused by high lipid levels and that, in turn, leads to plaque formation. Nitric oxide formation is highly reduced by the effect of saturated fats and oils on the endothelium of cells. Two hours after eating a fatty meal there is a significant drop of nitric oxide production. It will take nearly 6 hours, in fact, for the endothelium function to get back to normal.[5]

If a single meal can have such an impact on vascular health and cell membrane damage, then imagine the damage that is done by three meals a day, seven days a week, and 365 days a year –for decades! No wonder people die of heart-related diseases and other chronic diseases. And remember that here we are talking only about high saturated fats, cholesterol and oil diets, especially diets consisting of animal products.[6]

So, is there any point in eating fruits and vegetables with a piece of steak or chicken breast and letting yourself believe that you are eating healthily? Is it healthy to eat a sandwich consisting of white bread spread with margarine, a slice of cheese, topped-up with some lettuce, onions and tomatoes? You should not go half way when you are not feeling well or whenever you want to improve your health. It is that simple!

When it comes to nutritional excellence you need to eat the healthiest foods every day, whenever possible. There cannot be half-way houses. A plant-based diet and lifestyle change is essential for 100% health and for increasing human longevity.

The problem in this modern era is that the foods that we have been taught and thought we healthy for us, are actually unhealthy and do us considerable harm. Most people have been brought up to think cow's milk and other animal products are good for their health. But today there is ample medical evidence that such foods are unhealthy. Yet, how many people have eliminated animal products from their diets? Not many! How many people still drink soda drinks or eat early morning breakfast cereals, refined and processed foods on a daily basis? You see what this is leading to? You will be surprised how many of these unhealthy foods you actually eat on a daily basis: long-term consumption of animal produce and junk-foods containing harmful chemicals can build-up free radicals, toxicity and can cause havoc and disease to the human body.

Epidemiological studies, particularly those like EPIC, carried out on large populations, can offer tremendous insight into crucial public health questions, such as what we should eat and what we should not eat to minimise our risk of falling prey to the chronic diseases currently plaguing the world.

The golden standard is the international standard, where people are put on a certain diet and monitored. It is easy to make people make small changes, especially if you pay them. Many studies have shown that is relatively easy to persuade people to make small changes like adding grapes or nuts and seeds to their daily diet – especially if what you want them to do is to eat.

But there is mounting evidence showing that to achieve big changes in our health we need to make big, not small, changes in our diet. **Moderation kills!** "If you want to lower you risk, sure, you can eat in moderation, but if you want to eliminate your risk, or reverse a disease, you really have to take healthy eating seriously."[7]

People with chronic diseases will never get better by just eating "moderate" amounts of meat and dairy products, but they can control their condition. If you eat a whole rich plant-based diet and no refined, processed or animal-based foods, your overall risk of disease will be controlled, reversed or maybe even be cured.

Do not leave your decision to eat a healthy plant-based diet to the last; especially if you have been diagnosed with a serious chronic condition. More sensibly, we must all eat the best foods possible, to help reduce the risk of developing a disease, as once you have the disease it is harder to reverse or cure it, even with a plant-based diet. No other preventive strategy will help you more than eating the correct plant-based foods. Prevention is better than finding the cure.

The Scientific Evidence and Final Conclusions

So let us look at what the scientific studies and evidence has taught us. Why and how can this book dramatically improve your health and tackle the fight against most common westernized diseases, including breast cancer?

To summarize:

1. Humans may be considered evolutionally herbivorous – anatomically designed and physiologically structured to eat plant vegetation.
2. Cultures around the world that are less exposed to industrial pollutants and thus eat natural healthy foods, are less likely to develop cancer or other degenerative diseases.
3. We should stay away and be aware of so-called "natural foods" that may also contain harmful carcinogenic chemicals, artificial colours, food additives and preservatives.
4. Even though some supplements can help re-build the body's defences and fight cancer, nothing works better than a synergic approach in which thousands of abundant nutrients and phytochemicals contained in a plant-based foods work in synergy together.
5. Cow's milk is healthy only for baby cows. Human breast milk is best – and only until weaning time. Dairy products are detrimental to human health.
6. Animal products can increase human risk of cancer.
7. Thousands of phytochemicals and antioxidants found in plant foods have many anti-cancer fighting properties.

8. Fibre, found only in plant food is an essential food component: owing to its function in binding to carcinogens , controls glucose and insulin surges, helps to eliminate excess oestrogen and cholesterol from the human body, etc.
9. A natural medicinal plant food that helps against cancer, but will not make billion dollar profits to the cancer industry; and, is highly improbable it will not get patent for the usage and intervention treatment for cancer.
10. Animal protein increases cancer risk by increasing IFG-1 and the promotion stage of cancer (cf The China Study).
11. Many plant foods have angiogenesis properties that help reduce blood vessels formation: the reduction and aiding in the reversal of invasive tumour cells metastasis.
12. It is very important to perform a detoxification programme once or twice annually to help the body rid itself of toxicity, thus facilitating the natural tendency of the body to repair itself.
13. Sugar feeds cancer.
14. The cancer environment is dependent on the fermentation of sugar (glucose and fructose): strives on an anaerobic, acidic pH and oxygen-depleted environment.
15. Is important, especially for vegetarians, vegans and nursing mothers, to take daily supplementation doses of vitamins B_{12} and D_3.
16. Conventional medicine uses treatments that always damages the body, while alternative (natural) treatments consist in repairing the body's ability to regain control of its normal bodily functions and helps to fight off tumour cells.
17. PET scans and IPT are two superior diagnostic tools that help detect and kill cancer cells.
18. Rebounding is a form of exercise that helps lymphatic toxic drainage and boosts the body's immuno defences to help fight disease.

The evidence and truth is out there. I have learned from the best – from those clinical nutritionists and other alternative medical experts who fully understand the importance of lifestyle medicine for optimal human health. I have combined their knowledge and experience and placed this vital information in this book for all to follow. The ultimate healing powers of lifestyle medicine in absolutely fundamental and mandatory for the prevention and reversal of disease.

As Albert Einstein said:

"Nothing will benefit human health and increase the chances for survival of life on Earth as much as the evolution to a vegetarian diet."

I bless you all with good health and the foods that Mother Nature has naturally provided: to help sustain and regenerate our bodies to perfect health. Eat well and stay healthy.

It is now all up to you. If you do follow the guidelines of this book your risk of developing any acute or chronic disease will significantly diminish. If you do not, then it is very likely that disease *will* one day revival, in your body and within your family members bodies, if ofcourse they too, continue to eat foods that are unhealthy and their lifestyles do not amend. In other words "disease follows disease!" Conventional treatments will never, ever completely cure cancer, but will only hide the symptoms of disease. Alternative treatments using lifestyle medicine will not only help towards reducing the symptoms of disease, but, in most cases, will overall resolve the underlying cause of any disease.

The bottom line? A plant-based diet that incorporates lifestyle medicine are the key beneficial and determinant health factors towards human health and longevity: aiding towards disease prevention, disease control and disease reversal.

Notes for chapter 31

1. Baker, S.M et al. (2003). *Detoxification and Healing, the key to Optimal Health*. Published by McGraw-Hill; 2nd edition, p.157. Bollinger, Ty. (July 16th, 2006). *Cancer: Step outside the Box*. Chapter 10: Spoiled Rotten, pp. 211-224. Published by Infinity 510 Squared Partners.
2. Ibid.
3. Greger, M. (2013). Biblical Daniel Fast tested. Retrieved from http://nutritionfacts.org/2013/01/24/biblical-daniel-fast-tested/
4. Greger, M. (2012). *Tightening the Bible belt*. Volume 10. Retrieved from http://nutritionfacts.org/video/tightening-the-bible-belt/
5. Esselstyn, C.B. (2001). Resolving the Coronary Artery Disease Epidemic through Plant-Based Nutrition: *Preventive Cardiology*, **4**, pp.171-177
6. Esselstyn, C.B. (2008). Prevent & Reverse Heart Disease. "The Revolutionary, Scientifically proven Nutrition-based cure." Chapter 5: Moderation Kills, pp.35-46. Published by Avery Trade; 1st edition.
7. Ibid.

Resources

Some recommended Alternative Cancer Treatments and Healthy Eating websites

Jane Plant
www.janeplant.com

Ty Bollinger, The Cancer Truth
www.cancertruth.net

Michael Greger
www.nutritionfacts.org

Fuhrman: How to Live, For Life
www.drfuhrman.com

The China Study Community
www.thechinastudy.com

The Gerson Institute
www.gerson.org

An Oasis of Healing
www.anoasisofhealing.com

Nathan Pritikin Longevity Center & Spa
www.pritikin.com

Neal Barnard
www.nealbarnard.org

Credence Publications
www.credence.org

Dr. McDougall's Medical Centre
www.drmcdougall.com

The Ornish Spectrum
www.ornishspectrum.com

Caldwell B. Esselstyn, Jr.
www.heartattackproof.com

Susan Silberstein
www.beatcancer.org

CANCERactive
http://www.canceractive.com/index.aspx

Natural News (Mike Adams, the Health Ranger)
http://www.naturalnews.com/About.html

The Burzynski Clinic, First, Do No Harm!
http://www.burzynskiclinic.com/

WEIL TM, Andrew Weil, M.D.
http://www.drweil.com/

Healing Cancer Naturally
http://www.healingcancernaturally.com/index.html

Cancer Natural Cure (Dr. Hoover N.D.)
http://www.cancernaturalcure.com/index.htm

Suppressed Cancer Therapies
http://www.whale.to/cancer/therapies.html

Chris Beat cancer
http://www.chrisbeatcancer.com/

Outsmart Your Cancer (Tanya Pierce)
http://outsmartyourcancer.com/

Hope4Cancer Institute
http://www.hope4cancer.com/

BUDWIG CENTRE
http://www.budwigcenter.com/

Mexican Cancer Clinics
http://www.mexicancancerclinics.com/

Dr. Leonard Coldwell
http://drleonardcoldwell.com/

FOODMATTERS (You are what you eat)
http://www.foodmatters.tv/articles-1/healing-cancer-naturally-a-holistic-approach

Alkalize For health (Cancer-Self Treatment)
http://www.alkalizeforhealth.net/cancerselftreatment.htm

Cancer-Free (Bill Henderson)
http://www.beating-cancer-gently.com/

Recommended Environmental, Food, Healthy living, and Animal Welfare DVD documentaries

Dying To Have Known
Journey To Find The Evidence To The Effectiveness of The Gerson Therapy

Fat, Sick & Nearly Dead
With doctors and conventional medicines unable to help long-term, Joe turns to the only option left, the body's ability to heal itself. He trades in the junk food and hits the road with juicer and generator in tow, vowing only to drink fresh fruit and vegetable juice for the next 60 days. Across 3,000 miles Joe has one goal in mind: to get off his pills and achieve a balanced lifestyle. Watch and learn the power of plant foods and juicing.

Food Matters
Prevent Illness, Reverse Disease and Maintain Optimal Health Naturally

Food, Inc.
You'll Never Look at Dinner the Same way Again (parental guidance is advised)

Earthlings
Nature, Animals, and Humankind – Make the Connection (parental guidance is advised)

Forks Over Knives: Warning
This Movie Could save Your Life!

Fast Food Nation
The Truth is hard to Swallow

Got The Facts On Milk? The Milk Documentary (2011)
Addressing truth, myth and all in-between, the film is a humorous yet shocking exposition that provokes serious thought about this everyday staple. The film raises questions about dairy's role in cancer, osteoporosis, weight gain, asthma, acne, early menstruation, and more. It covers the preponderance of lactose intolerance in communities of colour, and explains why dairy consumption is fraught with high stakes: political, economic, ethical and environmental considerations.

Healing Cancer From Inside Out
A follow-up to Eating, Mike Anderson's pioneering production that showed the standard American diet as the "biggest cause of disease, disabilities, and death in the U.S. today". In this film he is highly critical of the medical profession and its arsenal of toxic chemical treatments for cancer that are doing more harm than good. All the while, doctors are ignoring and even suppressing alternative practices featuring diet and lifestyle changes that have benefited many people.

Hungry For Change
This DVD exposes shocking secrets that the diet, weight loss and food industry do not want you to know about: deceptive strategies designed to keep you "coming back for more". Find out what is keeping you from having the body and health you deserve and how to escape the diet trap forever. Featuring interviews with bestselling health authors and leading medical experts plus real-life transformational stories with people who know what it's like to be sick and overweight.

Making The Connection
A new film which invites you on a journey, together with a chef, a farmer, an MP, an athlete, a dietician, and a poet. Explore an exciting

lifestyle which combines delicious, healthy food with tackling many of the global challenges facing us today. Will you make the connection and become part of the solution?

Meat The Truth (2008)
A documentary which forms an addendum to earlier films on climate change. Although such films have raised public awareness about global warming, they have sadly missed out the impact due to intensive livestock production. Meat the truth draws more attention and demonstrates that livestock production and farming generate more greenhouse gases (methane) worldwide than all Planes, Buses, Lorries, Cars, Boats, Trains added together.

Simply Raw
Reversing Diabetes in 30 days: An independent documentary film that chronicles six Americans with diabetes who switch to a diet consisting entirely of vegan, organic, uncooked food in order to reverse disease without pharmaceutical medication. It worked!

Sicko
After exploring the predominance of violence in American culture in Bowling for Columbine and taking a critical look at the September 11th attacks in Fahrenheit 9/11, activist film-maker Michael Moore turns his attention to the topic of health care in the United States in this documentary that weighs the plight of the uninsured (and the insured who must deal with abuse from insurance companies) against the record-breaking profits of the pharmaceutical industry

Tapped
Examines the role of the bottled water industry and its effects on our health, climate change, pollution, and our reliance on oil. The documentary presents an overwhelming amount of evidence which will change the way anyone thinks about bottled and municipal water.

The Beautiful Truth
It has been said that more people live off cancer than die from it. This movie can put a stop to this travesty. Here is a very practical guide to the intensive nutritional treatment of cancer and other life-threatening

diseases that many would consider to have been impossible to obtain – thanks to the work of Max Gerson and his daughter Charlotte.

The Cove
Shallow Water, Deep Secret (parental guidance is advised)

The Gerson Miracle
Examines many of the elements of the Gerson Therapy, explaining why we are so ill and how we have in our grasp the power to recover our health without expensive, toxic or mutilating treatments, using the restorative forces of our own immune systems.

The 11Th Hour (2007)
A look at the state of the global environment, including visionary and practical solutions for restoring the planet's ecosystems. Navigated by Leonardo DiCaprio.

Cancer –The Forbidden Cures
Massimo Mazzucco, has done an extraordinary job, presenting a history that has long been hidden and falsified by Big Pharma and their minions. The suppression of many alternative cancer therapies. This DVD really opened my eyes to what is going on and what has been denied to people all in the name of profit.

Processed People
The manufactured food business is bigger than Big Oil; that kind of money buys inconceivably large amounts of propaganda, misinformation and corrupted science. Processed People is a wake-up call with factual, hard-hitting health commentary that is rarely heard. If you're searching for the un-processed truth about diet and health, look no further.

Super-Size Me
While examining the influence of the fast food industry, Morgan Spurlock personally explores the consequences on his health of a diet of solely McDonald's food for one month. Watch what happens.

Famous quotes about medicine, health, diet and animal welfare

Albert Einstein: Nothing will benefit human health and increase the chances for survival of life on Earth as much as the evolution to a vegetarian diet.

Leonardo Da Vinci: I have from an early age abjured the use of meat, and the time will come when men such as I will look upon the murder of animals as they now look upon the murder of men.

Charles Darwin: The love for all living creatures is the noblest attribute of man.

Thomas Edison: Non-violence leads to the highest ethics, which is the goal of all evolution. Until we stop harming all other living beings, we are still savages.

Thomas A Edison: The doctor of the future will give no medicine, but will interest his patient in the care of the human frame, in diet, and in the cause and prevention of disease.

Mohandas Gandhi: To my mind the life of a lamb is no less precious than that of a human being. I hold that the more helpless a creature the more entitled it is to protection by man from the cruelty of man.

Abraham Lincoln: I am in favour of animal rights as well as human rights. That is the way of a whole human being.

Alice Walker: The animals of the world exist for their own reasons. They were not made for humans any more than black people were made for whites, or women created for men.

George Bernard Shaw: Animals are my friends; I don't eat my friends.

Pythagoras: Animals share with us the privilege of having a soul.

Thomas Paine: Everything of persecution and revenge between man and man, and everything of cruelty to animals, is a violation of moral duty.

Henry Salt: The emancipation of men from cruelty and injustice will bring with it in due course the emancipation of animals also. The two reforms are inseparably connected, and neither can be fully realized alone.

Henry David Thoreau: One farmer says to me, "You cannot live on vegetable food solely, for it furnishes nothing to make bones with" and so he religiously devotes a part of his day to supplying his system with the raw material of bones, walking all the while he talks behind his oxen, which, with vegetable-made bones, jerk him and his lumbering plough along in spite of every obstacle.

Buddha: To keep the body in good health is a duty; otherwise we shall not be able to keep our mind strong and clear.

World Health Organization: Health is a state of complete physical, mental and social well-being, and not merely the absence of disease or infirmity.

Hippocrates (460-357 B.C.): He who does not know food, how can he understand the diseases of man?

Hippocrates: Let food be thy medicine and medicine be thy food.

Tiruvalluvar: How can one, who eats the flesh of others to swell the flesh, show compassion?

Helen Nearing: Flesh-eating by humans is unnecessary, irrational, anatomically unsound, unhealthy, unhygienic, uneconomic, unaesthetic, unkind and unethical. Need I elaborate?

Mike Adams: Today, more than 95% of all chronic diseases are caused by food choice, toxic food ingredients, nutritional deficiencies and lack of physical exercise.

Menander (342-291 B.C.): Health and intellect are the two blessings of life.

Dr. Thomas Fuller: Health is not valued till sickness comes.

Thomas Jefferson: If people let the government decide what foods they eat and what medicines they take, their bodies will soon be in as a sorry state as the souls who live under tyranny.

Charlotte Gerson: Modern allopathic medicine is the only major science stuck in the pre-Einstein era.

Louis Pasteur: Wine is the most healthful and most hygienic of beverages.

Dr. Dean Ornish: I don't understand why asking people to eat a well-balanced, vegetarian diet is considered drastic, while it is medically conservative to cut people open.

Genesis 1:29-30: Then God said, "Behold, I have given you every plant yielding seed that is on the surface of all the earth, and every tree which has fruit yielding seed; it shall be food for you; and to every beast of the earth and to every bird of the sky and to everything that moves on the earth which has life, I have given every green plant for food"; and it was so.

John Sammut: Scientific evidence has clearly proven and shown us that cultures that have not yet been westernized; that eat nature's plant foods, do live fruitful, longer and healthier lives. Let us be more sensible and follow their dietary habits to help sustain human health, animal welfare and for the future sustainability of this wonderful and fragile planet we call Earth.